SOLAR HEATED
BUILDINGS
of North America:

120 Outstanding Examples

SOLAR HEATED BUILDINGS

of North America:

120 Outstanding Examples

WILLIAM A. SHURCLIFF

BRICK HOUSE PUBLISHING CO.
Church Hill
Harrisville, New Hampshire 03450

Acknowledgments

Extensive help from more than one hundred solar architects, engineers, inventors, builders, and owners of solar houses has made this book possible. Without their help it would have been impossible to assemble so much up-to-date factual information.

Production Credits:

Book and Cover design: Diana Esterly
Typesetting and illustrations: G. Wilkins & Associates
Copy Editor: John Woodman
Editor: Jack D. Howell

Warning concerning patents and trademarks

Many of the components and systems used in the solar buildings described here are covered to some extent by patents or patent applications. To determine which matters are so covered, and which not, would be an almost impossible task and the writer has not attempted to make such determination. Persons considering reproducing components or systems described here should make independent determination as to whether patents are involved.

Some of the names of materials and systems have been trademarked. To determine which have been trademarked, and to include in each instance a warning that a trademark exists, would have been impractical.

CONTENTS

INTRODUCTION

This is a practical book, describing 120 existing buildings. Their solar heating systems, representing the best thinking of more than a hundred solar architects, engineers, and inventors, are described in detail. The 120 descriptions, with perspective drawings and photographs, emphasize practice, not theory. Little space is given to adjectives, or praise, or philosophizing.

The buildings chosen—houses, office buildings, schools, etc.—are representative of all parts of The United States—the East, Southwest, Midwest, South, Southwest, and Far West—and seven of the buildings are in Canada.

The descriptions are arranged by state, so that each reader can find easily what kinds of solar buildings have been built in his state and in neighboring states.

In selecting the 120 solar-heated buildings to be described here, I gave main attention to the solar heating systems themselves, this being a book about solar heating, not a book on architecture. The following criteria (questions) were kept in mind:

- Is the solar heating system effective? Does it provide a large fraction of the heat needed?

- Is it easy to operate?

- Is it durable?

- Is it cheap? More exactly, is it likely to be truly cost-effective? Does it furnish much heat per dollar of overall cost (first cost, operating cost, and maintenance cost)?

- Is it esthetically attractive? Is it generally compatible with a widely accepted type of building design?

- Is it of an unusual and technically interesting type that may now or soon show great promise?

Making the choice has not been easy, especially as there are so many types and sizes of buildings to be considered. There are houses, apartment buildings, office buildings, schools, churches, greenhouses, etc. Also there are one-story and multi-story buildings, buildings with horizontal roofs and buildings with steeply sloping roofs, buildings intended to be elegant and fully automatic and buildings intended to be extremely simple and controlled manually.

Some specific difficulties are:

- The performance of a solar-heated building is hard to judge, especially if, initially, the control system has not been adjusted properly, or if the occupants experiment with several different operating methods, or if hundreds of visitors troop through the building, opening and closing the outer doors and letting much heat escape.

- Imponderables such as comfort, convenience, esthetic appeal, and safety are important but hard to quantify.

- The durability of a system can scarcely be judged after only one or two years of operation—and many of the buildings are not yet two years old.

- Defining and evaluating the overall cost of a solar heating system is all but impossible when the system is the first of its type. Costs of manufacturing and installing the system may be unfairly high. Cost of planning may be unfairly low if the inventor makes no charge for his services. The cost of labor may be negligible if most of the work is done by the owner himself.

- The architects and owners who supply the performance data often wear rose-colored glasses.

There are certainly hundreds of excellent solar-heated buildings that are not described here. Limitations of time and publishing space prevented inclusion of more than 120 buildings. In particular, I have made *no* attempt to include solar-heated buildings that are merely of historic interest. Such buildings have been described in my 1977 book *Solar Heated Buildings: A Brief Survey,* 13th and final edition, which describes many of the historic solar-heated buildings here and abroad. Nor have I attempted to include buildings the solar heating systems of which are merely highly complex, merely highly publicized, or merely very costly. If I have any bias, it is in favor of solar heating systems that are simple, reliable, and inexpensive.

Why did I not concentrate on the *best* solar-heated houses? Because choosing the best is an almost impossible task. There are too many ways of defining *best* and too many difficulties in making the evaluations. Perhaps in five or ten years a body of architects may have the necessary information—and courage—to identify some *best* solar buildings. Very likely there will be many best types—perhaps twenty-five or fifty of them to suit different parts of the country, different types of buildings, different degrees of comfort and automation.

Great effort has been made to insure that the accounts are accurate and up to date. Most of them have been checked and rechecked with the architects, owners, etc. However, some errors may remain. No absolute reliance should be placed on any one fact presented unless it is confirmed independently.

TOPICS OF OUTSTANDING INTEREST

The variety of solar-heating-system designs is large and growing. Every month new types appear. Inventors are coming up with new and promising materials, new components, and new ways of combining old components. Their goals include better performance, reduced cost, greater durability.

* * * * * * *

Many new components and systems are described in this book. Some of the most interesting topics covered are listed below.

xii

water-filled tanks in which large quantities of fish are grown	275
water-filled plastic bags resting on a thin metallic horizontal roof	6
many concrete walls that are arranged like a series of vertical louvers and admit morning radiation and absorb afternoon radiation	250
a greenhouse as collector	1, 55, 82, 181, 188, 190, 228
a staircase-like roof that admits solar radiation in winter but not in summer	124, 129
a large array of Glauber's-salt-filled tiles, affixed to the ceiling of a room, that receive solar radiation via reflective, near-horizontal slats of venetian blinds	105

Among the special types of buildings described are

school and college buildings	47, 124, 252
office building	78, 175, 198
church	45
community building	20, 60, 149, 173, 213
arboretum building	80, 198
greenhouse	80
aquaculture building	275
very low-cost house	1, 28, 70, 82, 129, 178 188, 218, 233, 250, 261, 280

EXPLANATION OF SOME OF THE TERMS USED

Most of the terms used are clear enough, but a few deserve explanation.

Date The dates indicated are the dates of completion of the buildings and their solar heating systems.

Percent solar-heated This may be defined by an example. If the solar heating system of a given building contributes, in the course of a typical winter, so much heat that the amount of auxiliary heat needed (from oil furnace, or other conventional source) is reduced by 75% relative to a building that is identical except for having no solar heating, the given building is said to be 75% solar-heated.

(Warning: To judge a solar heating system just by the *percent solar heated* figure would, of course, be a great mistake. The figure depends on many circumstances, including size of building, extent of insulation, climate, and occupant's choice of thermostat setting.)

Slope of a roof or collector panel The angle between the roof or collector panel and a horizontal plane.

Passive solar heating system If a solar heating system does not employ machines for circulating liquid through pipes, or air through ducts, the system is said to be passive. Note that the system may, nevertheless, make use of shutters, dampers, vanes, or curtains that are operated at the start and finish of the day and may employ fans to help circulate air in an informal manner. A system that contains no mechanical moving parts whatsoever might be called an ultrapassive system; very few solar heating systems conform to this definition.

Selective coating A black coating that, while absorbing solar radiation strongly, emits little infrared radiation. Collectors employing such a coating perform especially well: much solar energy is collected and little energy is lost by radiation process. A typical selective coating has an absorptance for solar radiation of 90 to 95 percent of the theoretical maximum and has an emittance of 4-to-40-micron infrared radiation of about 10 to 30 percent of the theoretical maximum for an object of the given temperature. Absorptance and emittance are referred to as "a" and "e" respectively. An a/e ratio of about 9-to-1 is considered excellent. (Nonselective black coatings are very common and inexpensive, but the emittance may be as great as 90 percent and accordingly much energy may be lost by infrared radiation, especially if the surface is hotter than about 170°F.)

Beadwall A transparent or translucent window employing two glazing sheets a few inches apart, this space being filled, on cold nights, with tiny, insulating beads of polystyrene foam. During the day the beads are stored elsewhere and solar radiation is free to pass through the window. Beadwall was developed by Zomeworks Corporation.

Skylid An automatically actuated, solar-powered insulating system of louvers or shutters mounted close below a skylight. The device closes automatically when solar radiation ceases and opens automatically when solar radiation resumes. The device was developed by Zomeworks Corporation.

COP (coefficient of performance of a heat-pump) This is the ratio of *heat delivered* by the heat-pump to *electrical energy input* to the heat-pump. For example, if, during a typical day or week in winter, the given heat-pump delivers to the rooms 200 kWh of thermal energy, and if the total consumption of electrical energy by the heat-pump system as a whole is 100 kWh, the COP is 200 kWh/100 kWr = 2. The COP of a given heat-pump is, of course, variable; for example, it decreases when the difference between output temperature and input temperature is increased, or when operation is intermittent.

Absorption-type cooling system A system in which cooling is produced by the evaporation of water into a near-vacuum, the vacuum being maintained by absorption of the resulting water vapor by a liquid solution that contains much lithium bromide or other material with strong affinity for water. Solar energy can be used to recondition, or rejuvenate, the solution by heating it so hot that the water is driven off.

Chino Valley Solar Adobe Studio

Small, very low-cost, passively solar-heated studio with integral greenhouse and massive adobe walls

Chino Valley   34° N
(alt. 4800 ft.)
Star Route, Granite
Creek Lane,
½ mi E of
Highway 89

Building   This owner-built house, with main floor and loft, includes 550 square feet of living area and 150 square feet of greenhouse. The walls are of 12-inch-thick adobe with 1 inch of Styrofoam on the outer face, protected by ¾ inch of stucco cement. The greenhouse window area, 16 by 12 feet, is single-glazed with glass which is supported by steel rails, or channels, 4 feet apart. At night the window area is insulated by means of 4-by-4-foot shutters that are slid into place from above, guided and supported by the vertical steel rails. Each shutter consists of two 1-inch-thick Styrofoam panels in series, with ½-inch airspace and aluminum foil between.

Building: 1½-story, 700 sq. ft.	
Collector:	Passive
Storage:	
% Solar-heated: 70-80	

Passive Solar Heating System   Solar radiation enters the 200-square-foot greenhouse window and strikes the greenhouse growing area and also the massive walls of the building, which provide heat storage.

Auxiliary Heat Source   A wood-burning stove.

Domestic Hot Water   This is heated by a collector panel situated 10 feet west of the building.

Cooling in Summer   None. The greenhouse window area is covered with an "awning" made of snow-fence.

Designer, builder, owner: Michael Frerking of Equinox Company and William Otwell of Arizona Sunworks. *Cost:* The building was made almost entirely by hand (by three persons working three months) using adobe bricks made on-site and salvaged glass and lumber; the total expenditure for materials, other than the electrical system, was $1500.

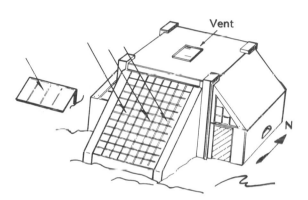

Flagstaff   35° N
(alt. 7000 ft.)
3450 Country Club
Drive, at Continental
Country Club

Building: 2-story, 2500 sq. ft.
Collector: 750 sq. ft., air type
Storage: 60-ton bin-of-stones
% Solar-heated: 75 (predicted)

Air-type solar heating system with 17-foot-high bin-of-stones

Building This wood-frame, saltbox-type house, with three bedrooms, 2½ bathrooms, attic, and two-car garage, has a floor area of 2500 square feet. There is an attic, but no basement. Ceilings are insulated with 6 inches of fiberglass. The foundations are insulated down to the footings. All windows are double-glazed. The living room and family room have cathedral ceilings.

Collection The collector is mounted on the 45-degree-sloping roof. The absorber is a ribbed sheet of nongalvanized steel with a nonselective black coating that was applied on-site. The glazing is single and consists of a sheet of double-strength glass. The space behind the black steel sheet is nominally 4 inches thick but contains many baffles that project 3 inches, which are intended to increase turbulence and heat transfer. The air is driven upward in this space by a ¾-hp blower situated in the attic. Behind the airspace there is a layer of foil-faced fiberglass batts.

Storage The 60 tons of 3-to-6-inch-diameter rounded granitic stones is housed in a tall, slender masonry bin forming the west half of the north wall of the main portion of the house. The bin is 16½ feet long by 4 feet wide by 17 feet high. It is strengthened by inclusion of many vertical and horizontal steel rods ½ inch in diameter. Hot air from the collector passes downward through the bin. When the rooms need heat, the furnace fan circulates room air upward through the bin and thence to the rooms via floor-level perimeter registers. Return air leaves the room via grilles at ceiling height. Automatically operated dampers are provided.

Auxiliary Heat Source A 165,000-Btu downdraft gas-fired furnace. Also Heatilator fireplace provided with two manually controlled fans that send hot air to the main rooms.

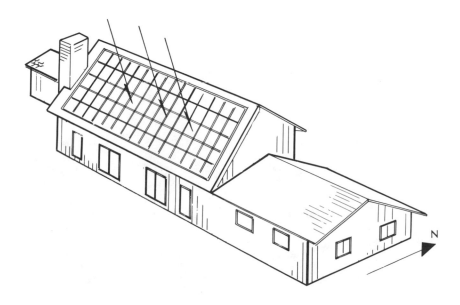

Domestic Hot Water   This is not heated by the solar heating system.

Cooling in Summer   None needed, none provided. Plans had been made for cooling the bin-of-stones at night by circulation of outdoor air and for using the bin to cool the rooms during the day, but these plans were dropped when it was found that no cooling was needed.

Problems and Modifications   During the second winter of operation several glass sheets of the collector broke, presumably because of strains resulting from slight shifting of the roof rafters. Also, the neoprene mounting strips for the glazing deteriorated. A new support system, employing redwood frames, was installed, and the replacement glazing sheets were of smaller size. Some relays in the fan control system failed and were replaced with relays having a higher rating. Initially the direction of airflow in the bin was the same for both hot input air from the collector and cool input air from the rooms—an arrangement which made thermal stratification in the bin a hindrance rather than a help. Specifically, the arrangement caused delay in warming the rooms when the thermostat setting was raised, unnecessarily prolonged operation of the pertinent blower, and precluded circulating hot air from the collector directly to the rooms. Changing to use of opposite directions of flow improved the performance greatly. The architect reports that in designing another such house he would use thicker glass (or tempered glass) on the collector and would use only half as great a mass of stones.

Solar engineer and architect: H. A. Wade. *Builder:*   H. E. Crain, Inc. *Assistance:* W. G. Delinger. *Owner:*   Yancy Lewis.

Sedona   35° N
(in center of state)
Red Rock Loop Road
(Route 2)

Building: 1-story, 1200 sq. ft.
Collector: 180 sq. ft., air type
Storage: 4 tons of stones
% Solar-heated: About 60

Retrofit air-type solar heating system

Building This wood-frame house, built in 1973 and equipped with solar heating in 1975, has three bedrooms, a small attic space, a built-in one-car garage, a 2-to-4-foot-high crawl space, and a shaded south porch. The windows are single-glazed. The house faces 10 degrees east of south.

Collection The heart of the collector, which is mounted on a roof that slopes 15 degrees, is a sheet of ribbed roofing aluminum painted nonselective black. Air, driven by a 1/5-hp blower, flows upward in the ½-inch space beneath this sheet. There are horizontal header ducts along the upper and lower edges of the collector. The glazing is a single sheet of Kalwall Sun-Lite, and there is a 2-inch dead-air space between glazing and black aluminum sheet.

Storage The 4 tons of 1½-inch-diameter stones is confined in a 1-ton concrete-block bin in the crawl space beneath the east end of the house. The rooms are heated either by direct circulation of hot air from the collector or by circulation of room air through the bin. In each case the same blower mentioned above is used. The control system, employing thermistor sensors and solid state electronics, operates six electrically powered dampers that direct the air flow.

Auxiliary Heat Source 7.5 kw electric baseboard heaters. Also a Heatilator fireplace.

Domestic Hot Water This is preheated in a small tank within the bin-of-stones. Final heating is by electric elements within a 30-gallon tank.

Cooling in Summer In the west end of the attic there is a 5000-cfm Arctic-Alpine evaporative cooler made by McGraw-Edison Company.

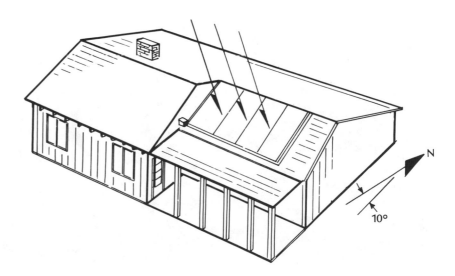

Employing moistened excelsior pads, a small water-feed pump, and a ½-hp blower, it delivers cool, slightly humidified air directly to the rooms.

Problems and Modifications   Initially the airflow space behind the black absorber sheet was 2 inches. This was later reduced to ½ inch in order to increase the linear speed of airflow along the absorber sheet.

Designer, builder, owner, occupant: R. G. Wagoner. *Funding:*   Private.

Atascadero   35½ °N
(midway between
Los Angeles and San
Francisco)
7985 Santa Rosa Road

Building: 1-story, 1140 sq. ft.
Collector: }
Storage: } Passive
% Solar-heated: 100

Harold Hay's world-famous house is 100% heated and cooled by a passive system that employs water-filled plastic bags on roof

Building   This is a one-story, split-level house, 38 by 32 feet. There are seven rooms, but no basement or attic. At the north end there is a patio and carport. The outer walls and partitions are of concrete blocks; the blocks of the partition walls are sand-filled, and many blocks of the north wall are vermiculite-filled. The solar heating system employs a combined collector and storage system on the horizontal roof; the roof serves also as heat distribution system and (in summer) heat dissipation system.

Passive Solar Heating System   On sunny days in winter, solar radiation strikes a set of side-by-side, water-filled bags covering practically the entire roof of the house (but not the carport roof). There are four bags, each 38 by 8 feet, with the long axes running north-south. They lie in the four alleys, or bays, between five spaced north-south tracks discussed below. Each bag is of transparent .020-inch polyvinyl chloride (PVC) sheet. The depth of water in each bag is 9 inches, and the total quantity of water is 6300 gallons (26 tons). No antifreeze is used; the water remains warm all winter. Above each bag there is a transparent sheet of ultraviolet-resistant PVC that is sealed along the edges of the bags and is held—by gentle air pressure from below—about 2 inches above the bags, thus providing some thermal insulation. Beneath each bag there is a nonselective-black sheet of polyvinyl chloride or polyethylene which rests on the metallic ceiling of the rooms below. The ceiling consists of 12-foot spans of ribbed sheet steel.

     Heat flows downward through the thin metallic ceilings into the rooms. Much of the heating is by radiation. The heating is sufficiently uniform throughout the house that no fans are required. The bathroom is heated by conduction through the walls and convection of air through the doorway. The thermal capacity of the massive walls and partitions supplements that of the 26 tons of water.

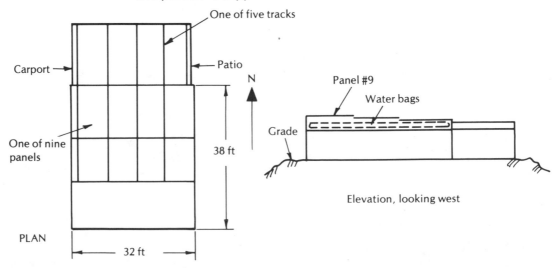

Diagram shows winter-night configuration, with the nine insulating panels in place above the water bags. (During day, panels are stacked above carport and patio.)

Throughout winter nights and summer days the bags are covered by a set of nine large, rectangular, horizontal panels of 2-inch-thick polyurethane foam. Panels 1 through 8 inclusive are 12 by 8 feet, and panel 9 is 33 by 12 feet. Each panel is supported by a set of ¼-inch-diameter, nylon-rimmed, ball-bearing-type wheels which run on horizontal, north-south, positive-guide rails of extruded aluminum. The extrusions are specially designed to provide three levels of wheel runway, to facilitate stacking.

During sunny days the panels are stacked, three deep, just above the carport at the north end of the building. The panels are moved automatically by means of a ¼-hp motor and sprockets and chains; these propel panel 9 (the large, southmost panel) which in turn propels panels 5, 6, 7, and 8; and these in turn propel the northmost panels 1, 2, 3, and 4. Alternatively, the panels may easily be moved manually, since they are lightweight. Seals prevent cold air from blowing between the panels and the bags.

In winter the panels are closed automatically whenever the energy-collection conditions (intensity of radiation and outdoor temperature) become unfavorable, as determined by a small model collector-dissipator consisting of an air-insulated horizontal black plate.

In summary, the water-filled bags collect solar radiation on sunny days, avoid losing energy to the outdoors at night (thanks to

the insulating covers), and continuously distribute heat downward, through the thin metallic ceilings, into the rooms.

Auxiliary Heat Source   None.

Domestic Hot Water This is not solar-heated.

Cooling in Summer   During the daytime in summer the water-filled bags are covered by the insulating panels and gain very little energy. At night the covers are moved aside and stacked above the carport so that the bags are exposed to the sky and to the cool night air and lose energy by convection and radiation. The pressure that keeps the uppermost plastic sheet a few inches above the bags is discontinued in summer, with the result that there is no insulating airfilm and the rate of energy loss is high. The bags remain cool throughout the summer; they cool the metallic roof which in turn cools the rooms. The extent of passive cooling is 100 percent.

Problems and Modifications   Initially there were some small leaks in the water-bag-and-roof system. These were corrected and there have been no leaks subsequently. Several minor problems arose in connection with the large horizontal insulating panels: problems concerning edge seals, thermal losses through frame members, drive system, and controls. These have been corrected. New designs of panel systems have been proposed recently.

Inventor and solar engineer: H. R. Hay of Skytherm Processes and Engineering. *Supplementary solar engineering:*   P. W. B. Niles. *Architect:*   Kenneth L. Haggard and John Edmisten. *Performance monitoring:*   By Kenneth L. Haggard and others from California Polytechnic State University at San Luis Obispo, with $42,000 funding by HUD. *Others assisting:*   Many persons from California Polytechnic State University at San Luis Obispo. *Cost of building and passive heating and cooling system:*   Reports by K. L. Haggard indicate that if certain economies apparent by hindsight had been made, the cost might have been as low as $27,500, and the incremental cost relative to a conventionally heated and cooled building might be of the order of a few thousand dollars or less.

Seven tall, water-filled tanks in sitting room collect and store solar energy

Building This square wood-frame house has three bedrooms, a small greenhouse, and a small loft space. There is no basement or attic. The house faces exactly south. In the southeast part of the building there is a two-story-high sitting room and in the southwest part are a greenhouse and a dining area. The walls and roof are insulated to R-19 and R-30 respectively. The edges of the 4-inch-thick concrete floor slab are insulated with 1 inch of Styrofoam.

Passive Solar Heating System Much solar radiation enters the vertical, south-facing, two-story-high window-wall serving the sitting room. The window-wall is 14 by 14 feet and is almost 200 square feet in area. It is double-glazed with glass. Besides providing the sitting room with daylight, the window-wall serves the main storage system, which is a row of seven vertical, cylindrical, water-filled tanks. Each tank is a spirally corrugated, galvanized steel culvert, 14 feet high and 18 inches in diameter, with a nonselective dark brown coating. Its flanged base is bolted to a 12-inch-thick concrete pad integral with the 4-inch-thick concrete floor slab. The top of the tank is closed and sealed, but stands free. The tanks are 24 inches apart on centers, and there are 6-inch spaces between them to permit entrance of light, to provide view of outdoors, and to insure that, at least throughout the period from 9:00 a.m. until 3:00 p.m., no tank shades a neighboring tank. The tank centerlines are 17 inches from the big south

Davis 38° N
(a 2800-degree-day site)
(65 mi NE of San Francisco)

Building: 2-story, 1620 sq. ft.
Collector: ⎱
Storage: ⎰ Passive
% Solar-heated: 80-90

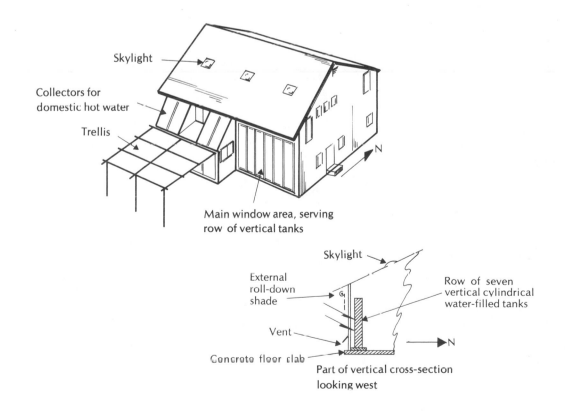

Skylight

Collectors for domestic hot water

Trellis

N

Main window area, serving row of vertical tanks

Skylight

External roll-down shade

Row of seven vertical cylindrical water-filled tanks

Vent

N

Concrete floor slab

Part of vertical cross-section looking west

window-wall, leaving an 8-inch space between tanks and that wall to provide adequate access for cleaning the windows. Each tank contains 24 cubic feet (1500 pounds) of water, and the set of seven tanks holds a little over 5 tons. Corrosion inhibitor (potassium dichromate) and algicide (copper sulfate) have been added to the water.

Additional solar radiation enters the west portion of the south face of the building, where there is a 125-square-foot area of vertical, double-glazed windows serving the greenhouse and the kitchen and dining area. This radiation contributes to the passive solar heating of the building.

When, on a sunny day, the upper story tends to become too hot, the hottest air—that is, air near the exposed-truss second-story ceiling—is collected and delivered, via a duct, to the colder first-story north room. For this purpose a small, manually controlled blower is used.

Auxiliary Heat Source   A 22,000-Btu/hr gas heater is used, and also a wood-burning steel Franklin stove situated in the sitting room close to the row of vertical cylindrical tanks.

Domestic Hot Water   This is preheated by a special water-type passive solar heating system situated in the upper southwest part of

the building. Solar radiation is received via four large, sloping windows comprising the roof of the greenhouse and dining area. These windows, which slope 45 degrees, are double-glazed with tempered glass. Their total area is 80 square feet. The radiation is absorbed by a set of six black-painted tanks installed close behind these windows. Each tank is a 33-gallon, cylindrical, glass-lined steel tank, 5 feet long and 1 foot in diameter, such as is used as a water tank in a recreational vehicle. The tanks are connected so as to form two groups of three. Within each group the tanks are hydraulically in parallel and the two groups are in series. Line pressure, assisted by gravity-convective forces, drives the water along in the system whenever a hot-water faucet is opened. No pump is used. After passing through this set of tanks, the water passes into a 20-gallon tank where a gas heater adds heat if this is necessary.

Cooling in Summer Although no formal cooling is provided, room temperature seldom rises above 78°F. Little heat enters the building, thanks to the well-insulated walls and ceilings, the eaves, the external roll-down shades that shield the large southeast window-wall, and the deciduous vines that cover the trellis adjacent to the southwest windows. The massive vertical tanks and the floor slab keep the temperature changes small. Hot air in the upper part of the building is allowed to escape via vents (operable skylights) in the roof. Cool air—often as cold as 55°F even on midsummer nights—enters via operable ports at the base of the window-wall adjacent to the row of vertical tanks. The air in the sitting room is kept in motion by a Casablanca-type fan mounted close to the ceiling.

Designer: Living Systems (especially Gregory Acker and M. B. Hunt). *Builders, owners, and occupants:* M. B. Hunt and Virginia Thigpen. *Cost of building as a whole, not including labor by owners:*   About $36,000. *Cost of solar heating system:* About $1700. *Funding:*   Private.

Gaviota House

Ten-zome house employing many skylights with automatic shutters

Gaviota 34½° N
(100 mi NW of Los
Angeles)

Building   The house is a cluster of ten "zomes" (modified domes with vertical walls) and has no attic or basement. Typical insulation is 4 inches of fiberglass in walls and ceilings. Interior walls are massive. South-facing windows total 150 square feet.

Collection   There are 300 square feet of skylights, immediately below which are Zomeworks Corp. Skylids—sets of thick insulating louvers that are automatically opened by a solar-powered mechanism whenever solar radiation becomes intense and are closed when solar radiation becomes negligible. The south windows also transmit some solar energy.

Storage   Heat is stored in the massive interior walls, floors, etc.

Auxiliary Heat Source   Four wood-burning stoves.

Domestic Hot Water   Special solar panels used.

Cooling in Summer   Not needed. The Skylids can be kept closed to exclude solar radiation. The large thermal capacity of the walls and floors prevents any sizable increase in temperature.

Solar engineer and architect: Zomeworks Corp., (Stephen C. Baer, R. Henry, et al.).

Building: 1-story, 2500 sq. ft.
Collector:
Storage: } Passive system; see text
% Solar-heated: 75 (predicted)

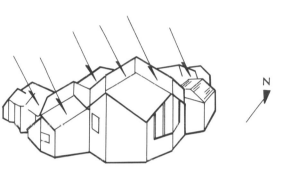

Hemet   34° N
(75 mi E of
Los Angeles)
Stanford Avenue at
Royal Palm Avenue

Buildings: 1-story, 1000-1500 sq. ft.
Collectors: 160-250 sq. ft., water type
Storage: 1000 gals. water
% Solar-heated: 80-90

A development consisting of 33 solar-heated houses using water type collectors

Buildings Of the seventeen buildings in the first set, all are Spanish-style stucco houses with two, three or four bedrooms, attached two-car garage, a small attic space, and no basement. Floor area varies with the number of bedrooms. The window area is from 10 to 15 percent of the floor area, and all windows are double-glazed. Extensive weatherstripping is used. The walls are insulated to R-11 with 4 inches of foil-faced fiberglass and contain a .006-inch vapor barrier; the ceilings are insulated to R-28. Many special energy-saving features are included.

Collection The collector varies in size depending on the size of the house, and is located, in most instances, on the horizontal roof of the garage. In the first set of houses, from nine to fourteen 6½-by-3-foot Revere Copper and Brass Company panels are hydraulically connected in series and parallel, with three panels in each series, with a slope of 44 degrees. The heart of each panel is a copper sheet with nonselective black coating, to which ¾-inch ID copper tubes are clipped and cemented at 6-inch intervals. The panel is double-glazed with 1/8-inch tempered glass and is backed with 2 inches of polyurethane foam. Water alone is used, and on cold nights freeze-up is avoided by circulating a small amount of water from the storage system. When the collector is operating, hot water flows directly from it into the storage system. In the second set of houses, completed in 1977, all collector panels are hydraulically in parallel, rather than series-parallel, and freeze-up on cold nights is avoided by

drain-down. Insulation on the piping in these houses is superior to that of the first set.

Storage   The precast concrete tank is insulated with 5 inches of Styrofoam and is buried beneath the utility room, adjacent to the kitchen. The rooms are heated by forced hot air.

Auxiliary Heat Source   Gas-fired domestic hot water heater; fireplace.

Domestic Hot Water   Water is preheated in a coil immersed in the storage tank. In the second set of houses, the domestic hot water system employs thermosiphon solar heating, so that the pumps used in space heating can be left off during the summer.

Cooling in Summer   Conventional cooling is used.

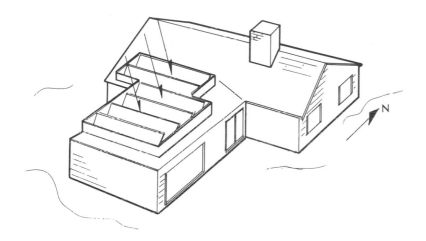

Problems and Modifications Early troubles included some leaks, some difficulties due to air in the pipes, and inadequacy of some check valves, which were replaced with foot valves.

Architect: W. F. MacDonald. *Developer and owner:* Blue Skies Radiant Homes, W. D. Buckmaster, President. *Builder:* M. Lauren. *Funding:* Private, by Bank of America. Little or no government funding involved. *Selling price:* $37,900 to $45,900, including solar heating system.

San Luis Obispo 35½° N
(midway between
Los Angeles and San
Francisco)

House heated by large, owner-built, very low-cost, water-type collector on hill near house, with storage provided by 50-ton concrete floor slab

Building   Solar heating was applied in 1973 to a house built in 1952. The house is 80 by 20 feet, with the long axis east-west. There is no basement or attic. The floor is a 6-inch-thick concrete slab resting on sand. The walls contain little insulation. The ceiling includes 1 inch of fiberboard beneath composition shingles. The windows, which are single-glazed, have a total area of 300 square feet. There are three double-glazed, 1-by-2-foot skylights. The porch roof, which extends 8 feet toward the south, unfortunately blocks much winter sunlight near midday.

Collection   The 800-square-foot water-type collector is on the south slope of a hill, 25 feet to the north of the house, and slopes 27 degrees. With overall dimensions of 40 by 20 feet, the collector consists of two 20-by-20-foot panels. Each employs a layer of nonselective black paint ("wrought iron paint") on a .009-inch aluminum sheet resting on a ¾-inch-thick, black sheet of fiberboard. These are backed by wooden boards. Immediately in front of the black aluminum sheet, lightly touching it, is an array of ½-inch-diameter black polyethylene tubes; the array consists of three separate loops connected in parallel to a 1¼-inch-diameter polyethylene

Building:	1-story, 1600 sq. ft.
Collector:	800 sq. ft., water type
Storage:	50-ton concrete floor slab
% Solar-heated:	75

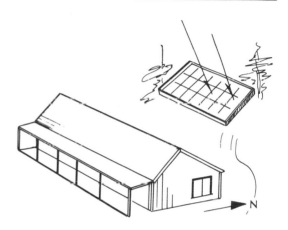

manifold. The aggregate length of tubing is 3600 feet. The tubing is held in place by guideposts in the form of 3-inch-long galvanized nails driven into the black fiberboard. Each panel has a 3-inch-high edge of wood. In the midwinter months the panels are single-glazed with a .006-inch sheet of clear polyethylene held 6 inches from the black fiberboard by an 8-by-5-foot mesh lattice of wooden strips, 2 inches by 1 inch in cross section, which are secured to the panel edges by staples and wooden battens. Above, there are confining strands of cord. Before mid-November, and also after mid-March, there is no glazing, so that the polyethylene tubing will not become so hot that it is damaged. Each year the old polyethylene glazing in discarded in the spring and new glazing is installed in the autumn. Installation takes 2 hours. Whenever the temperature of the collector exceeds 110°F, a ½-hp pump circulates plain water through the tubing and to the storage system.

Storage   Thermal storage is provided by the 80-by-20-foot, 6-inch-thick concrete floor slab of the house. Embedded in this 50-ton slab is a serpentine array of ½-inch-diameter copper tubing; the tube segments are 1 foot apart on centers. The total length of tubing is 1400 feet, and this is divided into four loops. Hot water from the collector is circulated directly through this tubing, and the rooms are heated by radiation and conduction from the floor. If, in sunny mild weather, the rooms threaten to become too hot, the water-circulation pump is turned off manually and, if necessary, windows are opened.

Auxiliary Heat Source   Propane gas heater.

Domestic Hot Water   About 15 percent of the above-described collector is allocated to heating this water.

Problems and Modifications   A small leak soon developed in a pipe connection and repair was made. The system has performed well for over four years, but it would have performed better if the copper tube segments in the concrete floor slab had been placed closer together. The performance would be more impressive if the house were better insulated. Before 1977 there was no aluminum sheet in the collector; this sheet was added to protect the fiberboard. Late in 1977 tests were being made to find whether adequate circulation of water could be maintained using a much less powerful circulation pump.

Solar engineer, owner, occupant: L. H. Carr. *Funding:* Private.

Santa Clara 37° N
969 Kiely Blvd.

Building: 1-story, 27,000 sq. ft.
Collector: Approx. 7000 sq. ft., water type
Storage: 10,000 gals water
% Solar-heated: 84 (predicted)

Large public building with a conventional water-type solar heating system

Building The center, located in Santa Clara's 52-acre Central Park, is a one-story building containing a great variety of large and small rooms, including meeting rooms and rooms for arts and crafts. The decision to employ solar heating and cooling was made after general planning was nearly complete, so that although the building was completed and first occupied in June 1975, the solar installation was not completed until April 1977. The outer walls of the structure are of concrete, 8 inches thick, without additional insulation.

Collection A total of 436 panels, each 8 by 2 feet, are located on four separate roof areas, three of which slope 18 degrees; the fourth is horizontal, and here the panels are tilted with respect to the roof and are also 18 degrees from horizontal. The panels, made by Ametek, Inc., are double-glazed and include Olin Brass Company Roll-Bond copper sheets with a selective black coating (a = 0.95; e ≤ 0.3) and integral passages for liquid. The insulating backing is of fiberglass and isocyanurate. The coolant used is water with 10% propylene glycol, with a maximum flowrate of 140 gpm. In summer, the collector outlet temperature reaches 220°F, sufficient to power the chillers.

Storage The hot water tank is a steel cylinder, 8 feet in diameter and 39 feet long, insulated with plastic foam and situated underground. The chilled water tank, 12 feet in diameter and 60 feet long, is also insulated with plastic foam and is underground. Rooms are heated by fan-coil systems.

Auxiliary Heat Source 1.2×10^{6} Btu/hr gas-fired boiler.

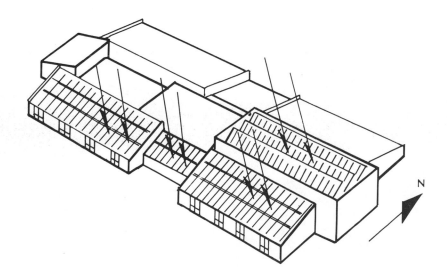

Cooling in Summer Two 25-ton Arkla LiBr absorption chillers are powered either by solar-heated hot water at 190° to 220°F or by the above-mentioned boiler. When demand is low, the chillers cool water in the chilled water tank. It is predicted that the building will be 65 percent solar-cooled.

Principal investigator: D. R. Von Raesfeld, City Manager. *Project director:* R. R. Mortenson, Director of Water and Sewer Utilities. *Scientific adviser:* J. N. Davis, of Licciardello, Davis & Associates. *Solar engineering:* Lockheed Palo Alto Research Laboratory. *Architect:* D. C. Thimgan. *General Contractor:* Welch Construction Company. *Others assisting:* Wilson, Jones, Morton & Lynch; University of Santa Clara. *Funding:* City of Santa Clara; ERDA; and American Public Power Association. *Cost:* Building: $1,010,000; Solar installation: $550,000.

The Sea Ranch 39° N
(100 mi N of
San Francisco)

Building: 1-story, 1150 sq. ft.
Collector: } Passive system—see text
Storage: }
% Solar-heated: Approx. 95

Passively solar-heated house, a large fraction of which is underground

Building The two-bedroom house, with loft, is of concrete and frame construction. Much of the living area is underground, opening onto a sunken patio at the south side. The house is sunk 4 feet into the ground and is surrounded on three sides by earth berms that join at the north with the sod roof. A detached garage and a studio are not solar-heated.

Collection There is a 375-square-foot double-glazed glass south window area, with a slope of 70 degrees, facing the patio. An insulating space-blanket curtain is used at night to prevent heat loss.

Storage Much heat is stored in the massive walls, etc., of the house.

Auxiliary Heat Source Wood stove.

Domestic Hot Water There is a 64-square-foot thermosiphon solar collector and an 80-gallon storage tank. No heat exchanger is used.

Cooling in Summer   Many vents are provided.

Problems and Modifications   During early operation, some reverse flow of hot water from the domestic hot water system to the collector occurred.

Solar engineers: David Wright, M. Chalom, K. Haggard. *Architect and owner:* David Wright.

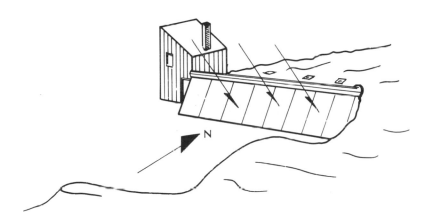

Walnut Creek 38° N
(30 mi NE of
San Francisco)
3199 Arbolado Drive

Building: 1-story, 2340 sq. ft.
Collector: 620 sq. ft., water type
Storage: 1500 gals. water
% Solar-heated: See text

A utility company's prototype solar-heated house employing a conventional water type system with a heat-pump

Building The wood-frame three-bedroom house is located in the Carriage Hills development. It has a small attic, an attached three-car garage, and no basement. The moderate window area is double-glazed, and the walls and ceilings are insulated to R-19 and R-30, respectively. Adjacent to the west portion of the house is a swimming pool that is heated by the solar heating system.

Collection On the south roof are thirty-four Revere Copper and Brass Company panels, each 6 feet 3 inches by 3 feet, mounted in two rows of seventeen panels each, at a slope of 18 degrees. Overall dimensions of the collector are 52 by 13 feet. Each panel consists of a copper sheet, with nonselective coating, to which are clipped rectangular-section copper tubes 5½ inches apart on centers. Panels are single-glazed with 1/8-inch tempered glass and are backed with 3½ inches of fiberglass. Coolant (water and antifreeze) is circulated at 25 gpm by means of a ½-hp centrifugal pump, and heat is delivered to the storage system by a heat exchanger.

Storage The horizontal cylindrical fiberglass tank, 5 feet in diameter, is insulated with 2 inches of plastic foam and is buried in the ground beneath the garage. A Tempmaster #200 2-ton water-to-water heat-pump is used to extract additional heat from the stored water. The house has two heating zones, each served by its own forced-air system.

Auxiliary Heat Source None, other than the above-mentioned heat-pump.

Percent Heated by combination of solar-heating system and heat-pump: 100.

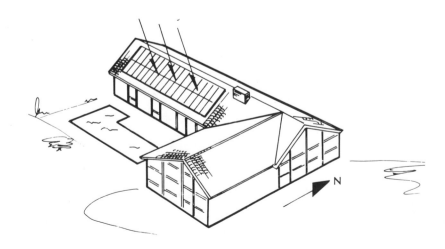

Domestic Hot Water Water is preheated in a coil in the hot-water storage tank and further heated by a gas heater.

Cooling in Summer Cold water is stored in a separate 1500-gallon tank and on clear summer nights is passed through the collector to be cooled further. Also, a heat-pump is used to dissipate heat to the swimming pool. To forestall a rise of the swimming pool's temperature over 90°F, which considerably reduces the efficiency of the heat-pump, the pool covers are removed and the pool "waterfall" is turned on. The pool is heated by the solar-heating system and by the heat-pump.

Problems and Modifications Soon after the collector panels were installed and were being readied for testing, they were inadvertently left filled with water during a cold weekend, and they froze and ruptured. They were replaced with new panels. After several months of use, the fiberglass hot-water storage tank developed a leak in a seam, which was repaired.

Solar consultant: Jack Schultz. *Builder:* Woodbridge Organization. *General supervision, funding, owner:* Pacific Gas & Electric Company.

Boulder   40° N
2000 Fifth Street

Building: 4-story, 5000 sq. ft.
Collector: 600 sq. ft., air type
Storage: 23-ton bin-of-stones
% Solar-heated: Approx. 50

Four-story condominium with air-type collector, bin-of-stones storage system.

Building   The building contains four apartments, one per floor, with plan dimensions of 44 feet (east-west) by 35 feet (north-south) each. There is no attic or basement. Cellulosic insulation is used: 6 inches thick in walls and 10 inches in the horizontal roof. The inner face of the first-floor concrete wall is insulated with 2 inches of polyurethane foam. Windows and external glass doors are double-glazed. Passive solar heating via windows is included in the estimate of 50 percent solar-heated.

Collection   The vertical collector is mounted on the south side of the building, with its lower edge about 15 feet from the ground. Windows intervene, dividing the collector into three areas. The collector is made up of a sheet of galvanized steel with a nonselective black coating, two layers of double-strength tempered glass in panes 2 feet 10 inches by 6 feet 8 inches, and a backing of 3 inches of fiberglass and 6 inches of cellulosic insulation. Two blowers circulate air upward through the collector in the 1-inch airspace behind the black sheet. A ⅓-hp blower compresses air entering the collector and a ¾-hp blower provides suction to move air out of the collector. The collector supply and return ducts each have a cross-section of 1 by 2 feet.

Storage   The bin, 10 by 6 feet by 7 feet high, is located in the southwest corner of the first story and is filled with 2-inch-diameter stones and insulated with 5 inches of polyurethane foam. Nominal pathlength in the bin is 6 feet. When an apartment needs heat, a ¼-hp blower circulates air from that apartment through the bin. Each apartment has its own thermostat and blower.

Auxiliary Heat Source   Each apartment has two in-duct electric heaters, rated at 2 kw and 4.8 kw.

Domestic Hot Water   Not heated by solar heating system.

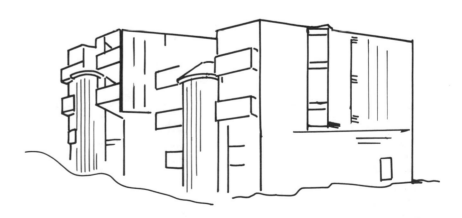

Cooling in Summer None needed; none provided.

Problems and Modifications Originally, the collector was served by a single blower. Various air leaks were discovered in the collection system: in the collector proper, in some of the duct attachments, and in one damper. To reduce pressures, and thus lessen air leakage, the developer installed the two-blower system described above.

Solar engineer: R. H. Bushnell. *Developer and contractor:* Kinetics Construction Corporation (R. White). *Funding:* Private

Denver   40° N
435 St. Paul Street

Building: 2-story, 800 sq. ft. plus loft
Collector: 200 sq. ft., air type
Storage: 12-ton bin-of-stones
% Solar-heated: No estimate available

Small prototype house designed for mass production and using air-type collector

Building   The house is intended to be suitable for use in row-house or cluster projects and is designed for "quick mass field factory production." The plan dimensions are 20 by 20 feet. The first floor contains the bedroom, a den, and a bathroom. The living area is situated on the second floor, which tends to be warmer than the first, and consists of a kitchen-dining-living room and a small greenhouse area. Above is a 200-square-foot loft reached by a pull-down stairway-ladder. There is no basement. There are earth berms on the east, north, and west side. There are few windows, most of them on the south side. All windows are fixed and double-glazed, and ventilation units serve each room. Insulation is with 6 inches of rockwool batts in walls, from 3 feet below grade to the roofline, and 12 inches of rockwool batts in the roof. The roof of the loft is shaped and waterproofed to hold a 6-inch pool of water to be used in an experiment involving solar radiation absorption in winter and cooling by evaporating in summer.

Collection   The collector has a slope of 50 degrees and is made up of ten Solaron Corporation panels, each 10 by 2 feet. The heart of the panel is a sheet of galvanized steel with a nonselective black coating. The panel is double-glazed with double-strength window glass and backed with 2 inches of fiberglass. Air, driven by a ¾-hp blower, travels through the 1-inch space between the sheet and the backing at a flowrate of 400 cfm. Air flows through the ten panels in parallel via 12-inch-diameter header ducts at the top and bottom of the collector. Air can be circulated directly to rooms or, if they are warm enough, to the storage system.

Storage   Located under the center of the house is a rectangular plywood bin, 6 by 6 feet by 6½ feet high, filled with 1-inch-diameter stones and insulated with 6 inches of fiberglass. The nominal pathlength of air in the bin is 5 feet. Room air is circulated through the bin by the fan-and-duct system.

Auxiliary Heat Source   Separate electric radiant heating system for each room.

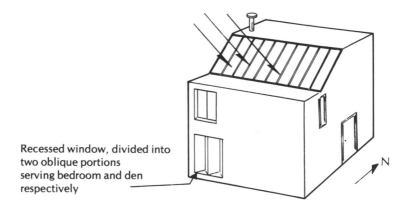

Recessed window, divided into
two oblique portions
serving bedroom and den
respectively

Domestic Hot Water Water is preheated by the solar heating system via a heat exchanger.

Cooling in Summer Natural venting of hot air from rooms is assisted in the afternoon by a "solar-heated chimney" that employs a vertical solar radiation panel affixed to the west wall of the house. Venting is assisted by a wind-driven rotating ventilator situated above the west end of the loft roof. The need for cooling is small because of the extensive use of insulation and the provision of a reflective blind for the west window.

Problems and Modifications The cover of the bin-of-stones warped, allowing leakage of air, but caulking sufficed to seal the leak. A temperature sensor was installed in the bin-of-stones, and a suitable control was added, to eliminate circulation of room air through the bin when the stones are cold. The motorized dampers of the air-handler failed to withstand the high temperatures of the airstream. New devices have been installed and have performed well. The ducts and apertures of the solar-heated chimney used for ventilation were of too small diameter, and the system performed poorly as a result. Plans are being made to improve the ventilation.

Architect: R. L. Crowther of Crowther Solar Group. *Builder:* Shaw Construction Company. *Owner:* D. J. Frey. *Cost:* House, $19,000; solar heating system, $3500.

Durango 37° N
(near SW corner
of state)
3112 E. Fourth Ave.

Building: 1½-story, 1775 sq. ft.
Collector: 260 sq. ft., air type
Storage: 8-ton bin-of-stones
% Solar-heated: No estimate

Well insulated house employing a small vertical air-type collector built on-site and containing a black fibrous absorber pad

Building   The wood-frame, three-bedroom house has an attached two-car garage and no basement. The first story includes a utility room, an unheated airlock entry, and an unheated food storage area. The attic is equipped with a turbine-type ventilator. The window area is small and all windows are double-glazed. Cellulosic insulation is used in the walls (5½ inches) and the ceiling (12 inches) and there is insulation around the perimeter of the first-story concrete floor slab as well as earth berms on the north and west sides. The building faces 15 degrees west of south.

Collection   The collector includes 15 vertical panels, built on-site, arranged in two groups. In each panel, the absorber consists of a 2-inch-thick pad of loosely and randomly arranged, inorganic fibers that have a nonselective black coating. The glazing consists of two layers of glass such as are used in double-glazed patio doors. Behind the fibrous pad there is an unpainted sheet of 26-gage galvanized steel and 1 inch Styrofoam. The fibrous pad is situated centrally in the 3½-inch space between the glazing and the galvanized steel sheet, and air passes transversely through the pad, picking up heat from it. The air is driven at 700 cfm by a blower situated in the utility room. Some solar energy is collected passively by the first-story view-windows, which have a combined area of 126 square feet, and by the 54-square-foot skylight in the first-story roof.

Storage   The 8 tons of 2-inch-diameter stones are contained in an insulated, rectangular, wood-frame bin in the utility room. The rooms are heated by a forced-air system that is controlled automatically and makes use of heat from the collector, the bin-of-stones, or the furnace.

Auxiliary Heat Source   Gas furnace.

Domestic Hot Water   Not heated by solar heating system. A gas heater is used.

Cooling in Summer   None needed; none provided.

Problems and Modifications   The initial performance of the system was poor. Malfunction of air-handler components caused incorrect airflow. Gaps at the edges of the storage system cover allowed hot air to escape. Repairs were made, and by late in 1977 performance was excellent. The owner believes that the storage system is somewhat too small and he may augment it with water in small containers.

Proposers of use of black fibrous absorber pad: Willie Heyen and Keith Olinger of Durango. *Architect and solar designer:*   B. B. Kesner. *Solar contractor:*   James M. Costello. *Owner and occupant:*   Harold L. Mansfield. *Funding:*   Private.

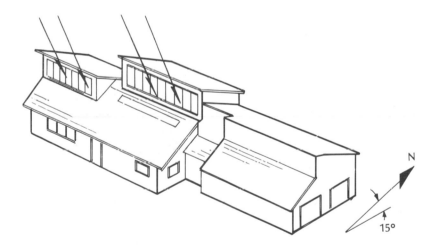

Meador House

Evergreen   40° N
(alt. 7400 ft.)
Hiwan Country Club

Building: 2-story, 2650 sq. ft.
Collector: 480 sq. ft., air type
Storage: 130,000-lb. bin-of-stones
% Solar-heated: 70

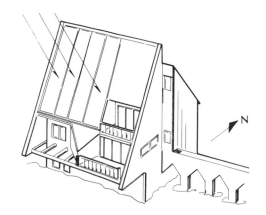

Tall house with large air-type collector and very large bin-of-stones

Building   The three-bedroom wood-frame house has a "basement" that is mostly above ground. The south-facing windows total 133 square feet, and some of them are deeply recessed to exclude direct sunlight in summer. The other three sides of the house have a smaller window area. All windows are double-glazed; the walls are insulated to R-18 with 3½ inches of fiberglass and an inch of Styrofoam; and the ceiling or roof is insulated to R-30 with 9½ inches of fiberglass.

Collection   The collector, located on the south face and sloping 53 degrees from horizontal, is made up of four panels designed and built by A-Sun-Do Manufacturing Company. Each panel is 24 by 5 feet and consists of a galvanized iron absorbing sheet covered with a nonselective black coating and double-glazed with Kalwall Sun-Lite. Behind the metal sheet is a 4-inch space, through which air is driven upward at 2200 cfm by a ½-hp blower.

Storage   The stones, about 4 inches in diameter, are contained in a rectangular bin, 12 by 10 feet and 11 feet high, located in the south-

east corner of the basement. This bin is constructed with a wood frame and Drywall and is insulated with 2 inches of Styrofoam board and 5½ inches of Styrofoam beads. When rooms need heat, room air is circulated upward through the bin by the above-mentioned blower.

Auxiliary Heat Source Gas furnace and fireplace.

Domestic Hot Water Water is preheated in a 100-foot coil in the upper part of the bin.

Cooling in Summer Little cooling is needed because of the 7400-foot altitude.

Problems and Modifications The dampers that control airflow did not close tightly at first and much heat was lost. The electrical control system did not work properly. Both of these matters have been corrected.

Architect, owner, occupant: Marlin R. Meador.

Fort Collins 40½° N
825 Strachan Drive

Building: 2-story, 1600 sq. ft.
Collector: 430 sq. ft., air type
Storage: 16,500-lb. bin-of-stones
% Solar-heated: 75

Compact, conventional-appearing house with air-type collector

Building The house is a two-story, three-bedroom Cape Cod salt-box, with an attic and crawl space. Insulation is provided by 4 inches of fiberglass in the walls and 10 inches in the ceiling directly below the roof, and the modest window area is double-glazed. A two-car garage is attached to the southeast corner.

Collection The air-type collector, 36 by 12 feet, is located on the south roof, which has a slope of 45 degrees. It consists of 54 panels, each 4 by 2 feet, manufactured by Solaron Corporation. A ⅓-hp blower circulates hot air from the collector either through the bin-of-stones or directly through the rooms.

Storage A bin 8 by 8 feet and 4 feet high, containing 1½-inch-diameter stones, is located in the crawl space. The bin walls are constructed of two sheets of ¾-inch plywood with 4 inches of Styrofoam between them. When the house is not being heated directly from the collector, room air may instead be circulated through the bin by means of a separate ⅓-hp blower.

Domestic Hot Water Water is preheated by passing it through a finned-coil heat exchanger mounted in a duct from the collector. There is an auxiliary gas heater.

Cooling in Summer An Essick Model BA 3000-2 evaporative cooler is used at night to cool the bin-of-stones.

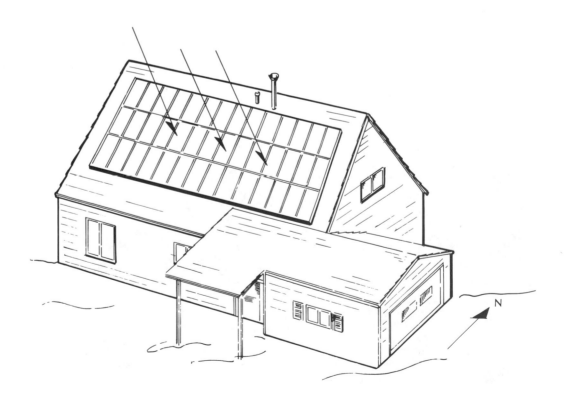

Problems and Modifications In three of the collector panels, it was necessary to replace the inner sheet of glass which had cracked. Inadequate duct insulation and duct sealing caused the second story to be too hot in summer. The insulation and sealing were improved and an automatic exhaust fan was installed in the attic. The cooling system did not operate properly during the first summer but has since been modified and is expected to perform well. The plumbing system of the solar domestic hot water heater was incorrectly designed for summer operation, necessitating use of the gas heater, but has been modified so that no summer use of gas is now required.

Architect: Eco-Era, Inc. (G. Frink et al.). *Solar heating:* Solaron Corporation. *Owner, occupant:* Roger E. Smith

Golden   40° N
(suburb of Denver)
Pine Ridge Road

Building: 2-story, 2500 sq. ft.
Collector: 292 sq. ft., air type
Storage: 7-ton bin-of-stones
% Solar-heated: 55 (predicted)

Air-type collector with vertical and sloping panels

Building   This is a two-story wood-frame house with four bedrooms. There is no basement or attic. The floor of the first story is a concrete slab 4 inches thick (16 inches thick beneath the bin-of-stones). Fiberglass insulation is 6 inches thick in the walls and 12 inches in the ceiling or roof. The house faces directly south, and windows on the south side are deeply recessed. Total window area is modest—about 10 percent of wall area—and all windows are double-glazed.

Collection   The collector is located on the south end of the main portion of the house. It employs fifteen Solaron panels, each 78 by 36 inches, in three rows of five panels; the lower two rows are vertical and the third is at a slope of 45 degrees. Air is driven upward in the collector, in the space behind the black absorbing sheet, at 600 cfm by a ⅓-hp blower incorporated in the Solaron driving-switching-controlling system housed in the utility room.

Storage   Beneath the collector is a rectangular bin, 8 by 5 feet and 7½ feet high, containing 1½-inch-diameter stones and insulated externally with 6 inches of Rockwool. Hot air from the collector is

driven downward through the bin. When rooms need heat, room air is circulated upward through the bin by the above-mentioned blower.

Auxiliary Heat Source   Electric baseboard heaters. Each room has its own thermostat.

Domestic Hot Water   Water is preheated by the solar heating system, by means of an air-to-water heat exchanger situated within the duct carrying hot air from the collector.

Cooling in Summer   None provided. The need for cooling is small, owing to the high altitude and the excellent insulation and deep recessing of the south windows.

Solar designer and architect: Crowther Solar Group (esp. R. L. Crowther). *Builder:* D. Peek. *Owner, occupant:* Murray Watts. *Cost of solar heating system:* $6000.

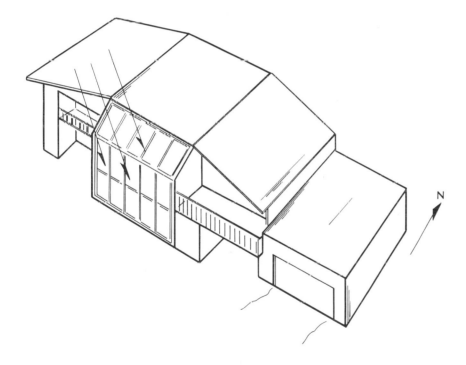

Gunnison 38° N
(alt. 8000 ft.)
(130 mi SW of
Denver)
Corner of Boulevard
Avenue and Denver
Avenue

Building	2-story, 3400 sq. ft.
Collector:	540 sq. ft., air type
Storage:	9-ton bin-of-stones
% Solar-heated:	Approx. 75

Air-type system used at very high altitude in very cold location

Building   The three-bedroom house has an attached three-car garage and no attic or basement. About 60 percent of the first-story floor area is a 4-inch-thick concrete slab. Excellent insulation is provided by 6 and 12 inches of urea formaldehyde foam in walls and ceiling or roof, respectively. The foundation, of cinder block with concrete footings, is insulated with 2 inches of Styrofoam. The modest window area is double-glazed. The house faces straight south, and most of the south windows are deeply recessed.

Collection   The collector is located on the south roof, which slopes 45 degrees. There are two horizontal rows of twelve panels each. The panels, each 78 by 36 inches, were made by Solaron Corporation and are similar to the Solaron panels described in accounts of other Colorado houses. The panels are double-glazed with 1/8-inch tempered glass. The air is driven upward along each column of two panels. The columns are in parallel, with circular header ducts 10 inches in diameter along the top and bottom of the collector. An airflow of 1000 cfm is maintained by a ⅓-hp blower situated above the storage system.

Storage   In the center of the first story is a rectangular plywood bin, 7¼ by 7⅓ feet by 8 feet high outside and 6¼ by 6¼ by 7½ feet inside, containing 3-inch-diameter stones. The bin is insulated with 6 inches of fiberglass. Hot air from the collector is driven downward through the bin, and when rooms need heat, room air is circulated upward through the bin by the above-mentioned blower.

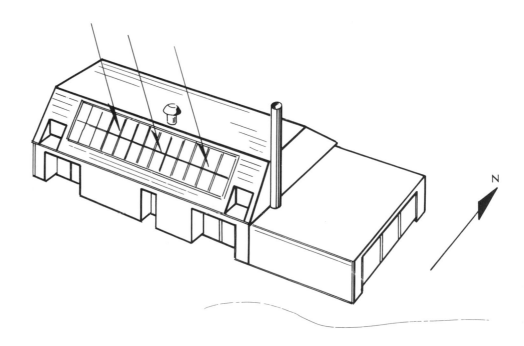

Auxiliary Heat Source   High-efficiency fireplace and electric base-board heaters.

Domestic Hot Water   Water is preheated by the solar heating system by means of an air-to-water heat exchanger.

Cooling in Summer   None, other than natural ventilation. Because of the high altitude and the good insulation and deeply recessed south windows, little cooling is needed.

Architect and solar designer: Crowther Solar Group (esp. R. L. Crowther). *Solar heating system supplier:*   Solaron Corp. *Contractor:*   G. R. Adelgren. *Owner, occupant:* Stanley Smock. *Cost of solar heating system:* $7000.

Montrose   38½° N
(in SW part of state)
Rolling Hills Estates

Building: 2-story, 2300 sq. ft.
Collector: }
Storage: } See text
% Solar-heated: 85 (predicted)

Small passive system teamed with a small air-type collector employing, as absorber, a batt of fine black fibers loosely and randomly arranged

Building The lower story of this wood-frame house is partly below grade, and the entrance is between stories. The house faces exactly south, and a two-car garage is attached to the west end. There is no basement or attic. The walls and ceilings are insulated to R-21 and R-40, respectively. Air leakage is greatly reduced by installation of a polyethylene film adjacent to the studs and trusses; the film is sealed with tape at all wall openings, such as those associated with pipes and electrical outlets. The below-grade portions of the walls are insulated on the outside with 2 inches of Styrofoam M, as is the below-grade concrete floor slab. The windows are triple-glazed. The exterior doors have foam cores and spring-brass weatherstripping. The overall heat requirement of the house is 8500 Btu per degree-day.

Solar Heating System There is a passive solar heating system and also an active solar heating system.

The passive system makes use of the large south windows, which have a combined area of 100 square feet.

The active solar heating system employs a 250-square-foot collector, 20 feet long and 13 feet high, situated on the garage roof, which slopes 60 degrees. There are two rows of collector panels, with seven panels per row, fourteen panels in all. Each panel is 76 by 34 inches. The panels were made by Tritec Solar Industries. The absorber consists of a 3-inch-thick batt of fine fibers that are loosely and randomly arranged and have a nonselective black coating. Air may pass through the batt easily. The panel is double-glazed with tempered glass; glass patio doors were used. There is an airspace in front of the batt and an airspace behind it. There are two input headers, one along each long vertical edge of the panel. The air issuing laterally (horizontally) from these headers turns and passes in a downward, north direction through the batt, picking up heat from it. This air then travels along beneath the batt to the output header at the upper end of the panel. (Note: The air that is traveling close beneath the glazing is air that has not yet passed through the batt and picked up heat, and because it is still relatively cool, the amount of

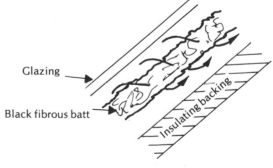

Glazing

Black fibrous batt

Cross-section of a portion of collector, looking west.

heat that is lost via the glazing is relatively small and collection efficiency is relatively high.) The air is driven at 700 cfm by a ½-hp blower. The storage system consists of a 160-cubic-foot bin containing 10 tons of 1½-inch-diameter stones. The bin, which is 8 by 5 feet by 4 feet high, is made of concrete and is insulated with 4 inches of Styrofoam. Air travels through the bin in horizontal direction. When the rooms need heat, room air is circulated through the bin at 700 cfm by the above-mentioned blower. Together, the passive and active systems provide an estimated 85 percent of the winter's heat need.

Auxiliary Heat Source   A 12-kw electric in-duct heater and a fireplace.

Domestic Hot Water   This is largely solar-heated by the above-described active solar heating system. The water passes along a coil within the bin-of-stones.

Cooling in Summer   None needed, none provided.

Solar engineer: R. H. Bushnell. *Architect:*   J. S. Lynch. *Solar equipment and installation:*   Tritec Solar Industries (J. M. Costello et al). *Contractor:*   Randy Brickson of Montrose. *Owner and occupant:*   R. Rusher and D. Rusher. *Funding:*   Private. *Cost of house as a whole:*   About $65,000. *Incremental cost of solar-heating system relative to conventional heating system:*   About $7000.

Shore House

Snowmass   39° N
(alt. 7200 ft.;
a 9200-degree-day
site)
(115 mi WSW of
Denver)

Building: 1½-story, 1450 sq. ft.
Collector: 564 sq. ft. (main), 48 sq. ft. (sup-
                        plementary)
Storage: 5300 gals. water
% Solar-heated: 100

High-altitude house with trickling-water collector and supplementary passive collector

Building   The first story (1050 square feet), which contains a kitchen-dining room, living room, studio, child's bedroom, and bath, is recessed 3 feet into the ground on the south side and 5 feet on the north side. The loft contains the master bedroom. The house is 38 feet east-west and 28 feet north-south and has a 12-foot-long greenhouse at the east end of the south wall. There is no basement. The house is well insulated and all windows are double-glazed.

Collection   The main collector is a trickling-water assembly located on the south roof, which has a slope of 50 degrees. It consists of fourteen panels, each 14 feet long and 2½ feet wide, containing two closely nesting sheets of ordinary .019-inch corrugated roofing aluminum. The upper sheet has a nonselective black coating. The collector is double-glazed with an outer sheet of double-strength glass and an inner sheet of single-strength glass, separated by a ½-inch spacer. The panes are supported by aluminum bars with a T cross section. Silicone sealant is used. Water is supplied to the collector by a ½-hp submersible pump (run at half its nominal power) and is introduced into the upper end of each panel via four tubes (one for every other valley of the lower sheet) and leaves via a single outlet tube at the bottom, at a flowrate of 10 gpm. The water contains no antifreeze or inhibitor and is drained into the storage tank before freeze-up can occur. The space between the aluminum sheets is so small (about 1/32 inch) that the water flowing down parallel to the corrugations wets a major fraction of each inner face, thus efficiently picking up heat from the irradiated upper sheet. This collector is operated only during the four or five coldest months of the year. In summer it is kept covered with a tarpaulin.

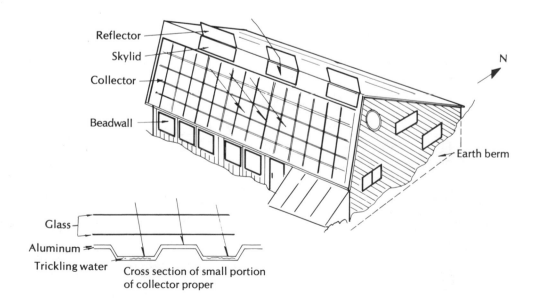

Reflector

Skylid

Collector

Beadwall

N

Earth berm

Glass

Aluminum

Trickling water

Cross section of small portion
of collector proper

The supplementary collector is located above the main collector. Near the ridge of the roof is a set of three skylights backed by Zomeworks insulating Skylids which open automatically at the beginning of a sunny day and close automatically when irradiation ceases. Each skylight is 2 by 8 feet, giving a combined area of 48 square feet. Three planar reflectors at the ridge of the roof, just above the skylights, with a slope of 62 degrees, direct additional radiation to the skylights, increasing their irradiation by a calculated 25 percent. Each reflector is 2 by 8 feet and is faced with .018-inch aluminized Mylar. In summer, the reflectors can be set horizontal in order to prevent direct radiation from reaching the skylights, but this has been found to be unnecessary. The double-glazed south windows beneath the main collector have a combined area of 90 square feet, of which about 60 square feet is Zomeworks-type Beadwall that is left clear on

sunny days and is closed at the end of the day by causing a blower to fill the 3-inch space between the glass plates with approximately 1 million 1/8-inch-diameter white polystyrene-foam beads, providing insulation comparable to 3½ inches of fiberglass.

Storage   The rectangular tank, 9½ by 12½ feet by 6 feet high, is of poured concrete, with walls 8 inches thick and the base 13 inches. It is waterproofed inside with a .03-inch layer of Plas-Chem 5500 butyl rubber applied with an airless spray gun and is insulated on the outside with 3 inches of polyurethane foam and 2 inches of Dow SM Styrofoam. The top is covered with 2 inches of insulation on which rests the particleboard floor of the child's bedroom. (This floor is 2 feet higher than the main ground-floor level.) Finally, the tank is surrounded by dry earth. Water from the main collector flows directly to the top of the tank; there is no heat exchanger. Water is fed to the collector from the bottom of the tank, where it is typically 10 Fahrenheit degrees colder than water at the top of the tank. The thermal capacity of the storage system is 44,000 Btu/°F (due to the water) and 8500 Btu/°F (due to the 308 cubic feet of concrete). The child's room is heated by energy leaking from the top of the tank. The other first-floor rooms are heated by radiant heat from the floor, which consists of a 3-inch layer of poured concrete in which are embedded a set of ¾-inch-diameter high-molecular-weight polyethylene pipes, 12 inches apart on centers. This floor rests on a 4-inch layer of poured concrete, which in turn rests on a 2-inch layer of Dow SM Styrofoam laid on earth. Circulation through the floor piping is usually by gravity convection, though a 1/12-hp centrifugal pump is sometimes used. An 80°F temperature of the circulated water suffices even in −10°F weather.

Auxiliary Heat Source   None.

Domestic Hot Water   During the months when the main collector is in use, water is heated in a loop of 1¼-inch-diameter high-molecular-weight polyethylene pipe, 100 feet long, running through the storage tank and delivering the water to a 52-gallon domestic hot water tank. Water flow in the loop is by thermosiphon only. In warmer months, a separate thermosiphon-type solar collector is used.

Cooling in Summer   None needed because of high altitude, good insulation, and high-thermal-capacity floor. None provided.

Problems and Modifications   Some of the glass sheets in the collector have cracked. The present sheets may be replaced with tempered glass. Some Tygon tubing used to supply water to the collector failed and was replaced with automobile heater hose. Although no significant corrosion of the aluminum sheets of the collector has occurred so far, the possibility of such corrosion is a source of worry to the owner.

Solar engineer, architect, builder, owner: Ron Shore. *Cost of solar heating system:* Approximately $3000 for materials, $500 for professional labor, plus considerable unpaid labor by owner and friends.

20,000 sq. ft. building nearly 100% heated by combination of 3100-sq.-ft. water-type collector and heat-pump

Westminster   40° N
(suburb of Denver)
9400 Wadsworth Blvd.
at 94th Street

Building A 17,300-square foot day-care center is joined to the 2700-square-foot church. There is no basement or attic. The church remains unheated five or six days a week, but on other days is given preferential heating while the day-care center is temporarily allowed to be cooler than normal. The walls and ceilings are insulated to R-17 and R-31, respectively. About half of the 1125-square-foot window area is south-facing. Single glazing is used throughout.

Building: 20,000 sq. ft. (church and day-care center)
Collector: 3100 sq. ft., water type
Storage: 14,000 gals. water
% Solar-heated: See text

Collection The collector is mounted on the large south face of the day-care center; this face slopes 55 degrees. The 82 panels, each 20 by 2 feet, are arranged in two rows. Each panel includes a sheet of 12-ounce copper that has a selective-black, copper-oxide coating by Enthone, Inc., with a and e values of 0.92 and 0.14 respectively. Copper tubes ¼ inch in diameter are silver-soldered to the copper sheet at 4-inch intervals. The coolant, which is water with no anti-freeze, is drained automatically before freeze-up can occur. The panels are single-glazed with 1/8-inch tempered glass and have a 2-inch insulating backing. The hot water from the collector flows directly into the storage tank; there is no heat exchanger.

Storage   Two water-filled tanks, of 8000-gallon and 6000-gallon capacity, are used. Each is a horizontal, cylindrical, steel tank and is insulated with fiberglass. They are situated below the collector. A water-to-water Carrier Hydromatic 50 MF/34 heat-pump is used also. It extracts heat from the smaller of the two water tanks and delivers it, via a small holding tank, to the forced-air distribution system. The heat-pump is used in this manner only; it is never used for cooling.

Percent Heated   by combination of solar heating system and heat-pump: 100.

Auxiliary Heat Source   None. There is no oil, gas or electric heater.

Domestic Hot Water   This is preheated by the solar heating system.

Cooling in Summer   None.

Problems and Modifications   Balancing and optimizing the controls took much time and care.

Solar heating system design, fabrication, and installation: R-M Products, Inc. (D. Erickson and others). *Architect:*   J. K. Abrams. *Funding:*   Private. *Cost of solar heating system:* $80,000

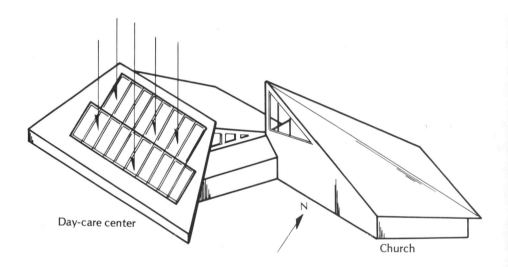

Day-care center

N

Church

Community College of Denver, North Campus

New college building with 1000-ft-long water-type collector and 200,000 gal. storage system

Westminster 40° N
(11 mi N of Denver)

Building The long, narrow building (1000 by 170 feet) is on a 160-acre site and occupies the crest of a long east-west ridge. Built to serve the needs of thirty-five hundred full-time students, it houses a variety of academic facilities. It is basically a rectangle, two stories in part and three stories in the remainder. There is excellent insulation (to R-17 in walls and R-12 in roof), and the windows are double-glazed. In addition, there are many energy-conserving and energy-redistribution features.

Collection The water-type collector has a gross area of 35,000 square feet and consists of two long east-west arrays that slope 55 degrees. The south array is mounted on a steel stand, and the north array is on a monitor that admits daylight and contains space for fans for forced ventilation. The collector panels were manufactured by Southwestern Sheet Metal Company. The heart of each panel is a

> **Building:** 2-3-story, 304,000 sq. ft.
> **Collector:** 30,000 sq. ft., water type
> **Storage:** 200,000 gal. water
> **% Solar-heated:** 100 (predicted)

sheet of 18-gauge steel to which seven ½-inch-diameter galvanized steel tubes are clipped, 5 inches apart on centers (see diagram). The sheet is coated with nonselective black 3M Nextel Black Velvet, and the panel is double-glazed with water-white glass. Water plus 50 percent ethylene glycol is used as the coolant, with a heat exchanger between the collector and the storage system.

Storage The water is contained in an insulated underground tank of poured concrete in two compartments of 100,000 gallons each. A chiller-heat-pump system is used to take heat from the storage system and deliver it, at a higher temperature, to multizone air-handling systems. The storage tank temperature may fall as low as 50°F and air at 110°F still can be delivered to the air-handler. When the storage tank temperature exceeds 110°F, water is delivered directly from the tank to the air-handler.

Auxiliary Heat Source   A gas-fired heater may be used.

Domestic Hot Water   In summer, water is heated by the solar heating system. In winter, the above-mentioned gas heater heats water as well.

Cooling in Summer   The chiller-heat-pump mentioned above is used, and rejected heat is sent to a conventional cooling tower.

Solar engineer: Bridgers & Paxton. *Architect:*   J. D. Anderson and Associates; many assisting groups. *Funding:*   By Colorado State Legislature, which in April 1974 appropriated the additional funds (about $700,000) needed to equip the building with solar heating. *Estimated cost of building:* $11,500,000 (including solar heating system).

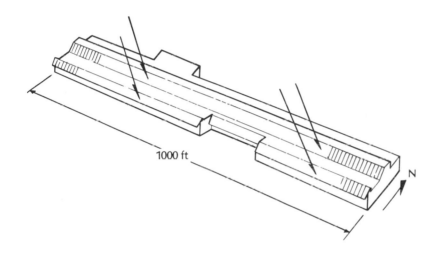

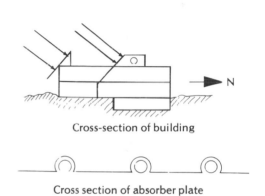

Cross-section of building

Cross section of absorber plate

Guilford   41° N
(12 mi E of
New Haven)

Building: 2-story, 1300 sq. ft.
Collector: 400 sq. ft., water type
Storage: Three parts; see text.
% Solar-heated: 60-70

1300-sq.-ft. house employing 400-sq.-ft. water-type collector and high-visibility 2000-gal. storage tank adjacent to living room. Backup storage is provided by subfloor bin-of-stones and by externally insulated concrete-block walls

Building   This square (28-by-28-foot) building has a clerestory and a loft. There is no basement. Below the floor slab is a 2-foot layer of 4-inch-diameter stones, below which there is a 2-inch insulating layer. The outer walls are of concrete blocks and their outer surfaces are insulated with 3 inches of sprayed-on urethane foam; thus the concrete blocks contribute to thermal storage and temperature stabilization. The windows are double-glazed. The north window area is small. The house includes many energy-conserving features. Some fireplace chimney heat is recovered and sent (by thermosiphon) to the storage tank.

Collection   The 400-square-foot, water-type collector, consisting of panels made by Sunworks, Incorporated, is on the roof, which slopes 57 degrees. A typical panel is 86 by 39 inches. The absorber is a .017-inch copper sheet, to the underside of which 3/8-inch-diameter copper tubes, 5 inches apart on centers, have been soldered. The selective black coating has an a/e ratio of about 10. The panel is single-glazed with a sheet of 3/16-inch glass. The panel has a fiberglass backing. The coolant, water to which 40 percent of ethylene glycol has been added, is circulated by a 1/12-hp centrifugal pump.

Storage   This system has three parts: (1) 2000 gallons of water in a vertical, cylindrical steel tank, 5 feet in diameter and 12 feet high, behind and just east of the fireplace. Much heat is received from the collector and some is received from the fireplace chimney; heat is distributed to the second-story rooms by a fan-coil system. (2) The 2-foot layer of stones below the floor. Heat is received on sunny days from warm air collected near the ceiling of the clerestory; later, heat is extracted from the stones by forced airflow. (3) The large quantity of concrete blocks of the exterior walls, which are insulated on the outside. The house has only one blower, which is rated at 1/3 hp at full power. On sunny days this blower drives air from top of the clerestory down to, and through, the 2-foot layer of stones beneath the floor. At night, this blower serves the very different purpose of driving hot air from within the jacket of the storage tank to, and through, the auxiliary heater (discussed below) and through the 2-foot-thick layer of stones and thence, via various openings in the periphery of the floor, into the rooms.

Auxiliary Heat Source   Oil-fired, 110,000 Btu/hr domestic hot water heater.

Domestic Hot Water   This is preheated by a separate 44-square-foot solar collector. Final heating is by the oil-fired heater mentioned above. In summer the main (400-square-foot) collector can be used to heat the domestic hot water, since there is then an excess of solar heat available.

Cooling in Summer The eaves shade the windows, and the south-side clerestory and belvedere encourage natural venting of hot air. The 2-foot layer of stones is cooled at night by forced circulation of outdoor air and can be used during the day to help cool the building. The main storage tank is left cool throughout the summer, which is permissible inasmuch as the domestic hot water system has its own tank.

Problems and Modifications The original heat exchanger in the bottom of the main storage tank was too small, having an area of only 20 square feet. It was replaced with a larger (70-square-foot) exchanger. Likewise the heat exchanger in the 65-gallon domestic hot

water tank was too small (10 square feet) and was replaced with a larger one (24 square feet). The steel inner wall of the main storage tank showed signs of corroding in the area of the air-water interface near the top when the tank was not full. The tank was drained, and its inner surface was sand-blasted and then coated with epoxy. Arrangements were made for supplying outdoor air directly to the oil burner and the oil-burner draft diverter, and (separately) to the fireplace, to avoid unnecessary loss of warm air. The fireplace heat-scavenging and thermosiphon system is being improved. The fireplace is being equipped with Pyrex doors. Plans call for installing thermal shutters on all windows at night and for installing a windmill (aerogenerator).

Solar engineer: E. M. Barber, Jr. *Architect:* C. W. Moore Associates. *Builder:* R. Riggio. *Owner and occupant:* E. M. Barber, Jr. *Contract amount:* $46,500.

Pinchot Guest House

420-sq.-ft. water-type collector provides 75% solar heating

Guilford   41° N
(12 mi E of
New Haven)

Building   This 32-by-24-foot, two-bedroom house has a basement, an unheated attic, and an attached three-car garage. The house walls are insulated with 3½ inches of fiberglass. The house faces 5 degrees east of south.

Building: 1-story, 750 sq. ft.
Collector: 420 sq. ft., water type
Storage: 1500 gals. water
% Solar-heated: 75

Collection   The 29-by-14½-foot collector, on the roof, which slopes 45 degrees, includes eighteen Sunworks, Incorporated, panels. Each is 86 by 39 inches and employs, as absorber, a .017-inch-thick copper sheet, to the underside of which copper tubes 3/8 inch in diameter and 5 inches apart on centers have been soldered. The black coating is moderately selective. The glazing is a single sheet of 3/16-inch tempered low-iron glass. The panel backing includes 3 inches of fiberglass. The aluminum frame also is thermally isolated. The supply and return pipes are of copper. The coolant is water with no anti-freeze; it is drained from the collector before freeze-up can occur.

The collector and pipes contain 25 gallons of water which is circulated at 7½ gpm by a centrifugal pump whenever the temperature of the collector exceeds that of the storage tank. Barber-Coleman controls are used. In summer, an air vent at the top of the collector is opened.

Storage   The rectangular steel storage tank, 10 by 4½ by 4½ feet, situated in the basement, is insulated with 6 inches of polystyrene foam. The rooms are heated by a fan-and-two-coil system, one coil being fed from the main storage tank and the other by the 30-gallon tank mentioned below. Each tank has its own circulation pump, and the appropriate pump is invoked automatically.

Auxiliary Heat Source   A Carlin oil-fired heater with a 30-gallon tank. This system comes into play when the temperature of the main storage tank is below 95°F. There is also a high-efficiency fireplace.

Domestic Hot Water   This is preheated by the solar heating system. Final heating is by the above-mentioned oil-fired heater.

Cooling in Summer   None.

Problems and Modifications   Initially, the supply and return pipes for the collector were of polyvinyl chloride. These were found to soften somewhat when at high temperature and were replaced by copper pipes.

Solar designer: E. M. Barber, Jr. *Architect:* Carleton Granbery. *Owner:*   G. B. Pinchot. *Funding:*   Private.

Partially underground, $40,000 house 70% solar-heated with the aid of several kinds of passive systems

New Milford   42° N
(30 mi NW of
New Haven)
110 Sawyer Hill Road

Building The house is 55 feet long and 22 feet wide. Situated on the west slope of a hill, it faces 5 degrees west of south. The first story, which is largely underground except at the west end, includes a family room, a greenhouse, and a storage room. The second story includes the living room, kitchen, dining area, and studio-bedroom. There is a dormer loft in the central upper part of the house. The walls are insulated externally with 3 inches of polystyrene foam water-proofed with white stucco cement. The roof contains 7 inches of fiberglass. The large south window areas, discussed below, are double-glazed and are equipped with shutters or heavy curtains that are closed on cold nights. The window areas on the east, north, and west are small.

Building: 2½-story, 2460 sq. ft.
Collector: } Passive
Storage: } Passive
% Solar-heated: 70 (predicted)

Passive Solar Heating System Radiation enters via the south windows, the total area of which is 500 square feet. Most of these windows (375 square feet) are vertical. Those serving the greenhouse in the southwest part of the first story (125 square feet) slope 60 degrees. All are double-glazed. Storage is provided by the massive

concrete walls and floors. The schemes used in the east and west ends of the building are very different. In the east end of the second story there is a vertical, south-facing, massive wall (Trombe wall) parallel to the window-wall and 2 feet from it; the 2-foot-wide space serves as a walkway. In the west end of the first story there is a 5-foot-high concrete-block wall about 5 feet from the underground south wall; the greenhouse occupies the 5-foot-wide space. The south faces of the Trombe wall and the 5-foot wall have been painted with a nonselective black paint, and they receive and absorb much solar radiation. Much diffuse sunlight passes over the 5-foot wall, illuminating the room north of it. The first-story portions of the east, north, and west walls of the building are of 10-inch-thick poured concrete, and the upper portions of these walls are of 10-inch-thick concrete blocks. The outer faces of these walls are insulated, as explained above, and the inner faces have been left bare, to facilitate intake and output of heat. Storage of heat in the poured-concrete first-story floor of the west part of the building is facilitated by a set of subfloor channels defined by north-south rows of hollow concrete blocks and served by east-west header ducts. Air can circulate through the channels by gravity convection (for heat output to the rooms), or a

fan can be used during heat input. The precast concrete panels that constitute the floor of the second story contribute to thermal storage.

Auxiliary Heat Source A fireplace and a wood-burning stove.

Domestic Hot Water This is preheated on passing along several parallel ¾-inch-diameter copper pipes in close contact with the above-mentioned Trombe wall. The pipes, which have a total length of 100 feet, are horizontal and are situated in a shallow recess, 1½ inch deep and 9 inches high, in the south face of that wall. The small storage tank is partly recessed in that wall. Final heating is by an electric heating element at the top of the tank.

Cooling in Summer There is no air conditioner. Because the building is well insulated and, in summer, the upper windows are shaded by roof overhang and the greenhouse windows are covered by canvas awnings, little cooling is needed. Openable windows and skylights vent hot air.

Solar designer, architect, builder, owner, occupant: Stephen Lasar. *Cost:* About $40,000. *Funding:* Private.

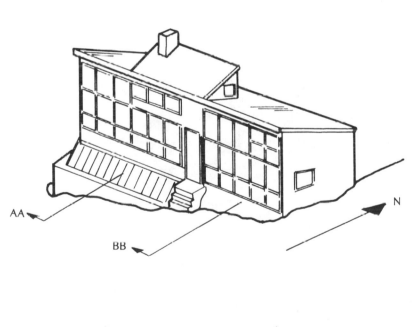

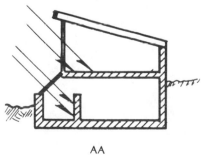

AA

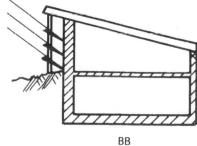

BB

New Milford   42° N·
(a 6700-degree-day
site)
(30 mi NW of
New Haven)
Squires Hill Road

Building	2-story, 1500 sq. ft.
Collector:	⎱
Storage:	⎰ Passive
% Solar-heated:	70 (predicted)

Passively solar-heated [70 % predicted] house that has 80 tons of walls and channeled floors

Building   The house is of wood and concrete and has two bedrooms. At the northeast corner there is an unheated, 510-square-foot attached two-car garage. There is no basement. The massive first-story floor consists of a 4-inch slab of concrete which rests on 8-inch-thick hollow concrete blocks laid on their sides to provide horizontal passages for air. The blocks rest on 2 inches of rigid insulation, which in turn rests on a 0.004-inch moistureproof film laid on gravel. The exterior walls are of concrete blocks insulated on the outside with 3 inches of Dryvit insulation with a waterproof coating. Below the frost line the outside is covered with 2 inches of rigid insulation. The roof is insulated with 6 inches of fiberglass. All windows are double-glazed. The window areas on the east, north, and west sides are small, aggregating 57 square feet. The south window area is large: 380 square feet. These windows employ Thermopane; some consists of Caradco assemblies and are openable, and others are sliding-glass-door assemblies that are fixed. The frames of the windows and doors are of insulating type. The front vestibule is of air-lock type. The masonry chimney is in the center of the house. The house faces exactly south.

Passive Solar Heating System   This employs 380 square feet of south-facing, vertical, double-glazed Thermopane window. Much of the solar energy received is absorbed directly or indirectly by the massive first-story floor slab and indirectly by the massive exterior walls, and some is absorbed by the channeled array of concrete blocks beneath the floor slab. The total mass of material absorbing energy is about 80 tons. When a thermostat in the south part of the second story finds the temperature there to be too high, it turns on the furnace fan, which draws second-story air downward (via a central vertical duct) and drives it through the above-mentioned subfloor channels and thence into the first-story rooms.

Auxiliary Heat Source   Oil-fired hot-air furnace of 105,000 Btu capacity. Also a Reginald wood-burning stove. The furnace sends some hot air to the second story, via ducts, and sends some hot air to the channels beneath the first-story floor slab.

Domestic Hot Water   This is preheated by a separate water-type thermosiphon solar heating system. The collector, employing 42 square feet of Sunworks copper absorber, is mounted at 40 degrees on the roof of a small extension (for entrance foyer and bathroom) near the southeast corner of the house. Hot water from the collector flows upward by gravity convection into a nearby 80-gallon tank situated on the second story. No antifreeze is used; the water is kept warm in sunless periods by gravity-convective circulation of room air through the space just below the collector. Final heating is electric, in a 40-gallon tank.

Cooling in Summer   The heat load is small because of the excellent insulation and the exclusion of solar radiation by eaves and awnings. The large thermal mass of the floors and walls helps reduce the temperature rise. The many openable windows provide cross-ventilation. There is no air conditioner.

Solar designer and architect: Stephen Lasar. *Builder:*   Design Associates. *Cost or price:*   Offered for sale by builder late in 1977 at about $84,500 including land, building, and solar heating system.

Garage window

Solar-Tec House

Old Lyme   41½°N
(in SE corner of state)
Wildwood Drive, at
Four Mile River
Landing

Building: 2-story, 1900 sq. ft.
Collector: 650 sq. ft., air type
Storage: 30-ton bin-of-stones
% Solar-heated: 65 (predicted)

House employing, as storage system, a slender 2½-story-high bin-of-stones in center of house

Building   This three-bedroom saltbox-type wood-frame house has a full basement and a small attic. The two-car detached garage at the north is reached by a covered walkway. The house walls are insulated with 3½ inches of fiberglass and 1 inch of Styrofoam. The ceiling or roof includes 5½ inches of fiberglass and 1 inch of Styrofoam. The window area is moderate: about 170 square feet in all. The windows are double-glazed. Insulating panels cover the sliding glass doors on cold nights. The house faces exactly south.

Collection   The collector, of special design and built on-site, is mounted on the south roof, which slopes 49 degrees. The collector is 34 feet long and 19 feet wide and includes thirty-six panels, each 6½ by 3 feet. The absorber is a sheet of 26-gauge galvanized steel with a nonselective black coating. The glazing is double and is of tempered glass. The panel backing includes 6 inches of fiberglass. Air flows in the 1½-inch space behind the absorber sheet; it travels upward in this space at 2000 cfm, driven by a ⅓-hp blower. The typical cross section of the supply and return ducts is 14 by 12 inches. The feeder ducts are 8 inches in diameter. The ducts are insulated with 1 inch of rigid fiberglass. Rho Sigma controls are used.

Storage   The 30 tons of 4-to-6-inch-diameter stones is contained in a slender vertical bin in the center of the house and adjacent to the west side of the chimney serving the central fireplace. The outside dimensions of the bin are 6 by 6 feet by 30 feet high and the inside dimensions are 5 by 3 feet by 29 feet high. The frame of the bin is of steel, and strengthening cross braces are used. The inner face is sheathed with steel sheets. The bin walls include 3½ inches of fiber-

glass, and on the outside there is a brick veneer. Air from the collector is driven downward through the bin at 1500 cfm by a ⅓-hp blower, and there is much resulting thermal stratification. When the rooms need heat, room air is circulated upward through the bin by a ¼-hp blower and is delivered to the rooms via several uninsulated ducts. Note that the bin performs four functions: it stores heat, it serves as the main duct for the collection system, it steadily leaks a small amount of heat into the rooms, and it picks up heat from the chimney whenever the fireplace is in use.

Auxiliary Heat Source An oil furnace and a Heatilator fireplace. The latter supplies heat to the rooms directly and also helps heat the bin-of-stones.

Domestic Hot Water Water from an 80-gallon tank passes through a finned coil within the upper part of the bin and is thus preheated. It then passes into a 52-gallon tank, where final heating is by electric heating elements.

Cooling in Summer There is no formal cooling system. An operable vent in the roof, near the chimney, allows hot air to escape. The bin-of-stones remains idle and cool in summer.

Problems and Modifications None.

Solar designer and builder: George C. Field Company (especially L. R. Clark). *Fabricator of the air handling system:* Bacon Brothers, Incorporated. *Owner and occupant:* A. Bradford. *Cost of house and solar heating system:* About $77,000. *Cost of solar heating system:* About $8000. *Funding:* Private.

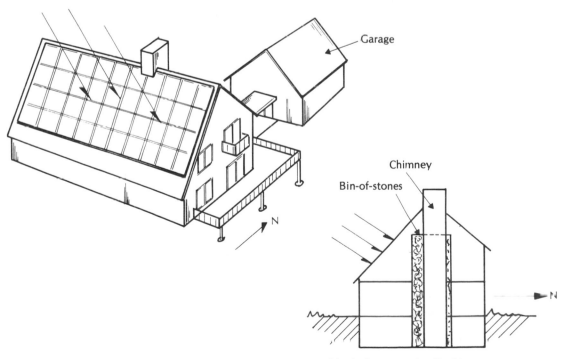

Vertical cross section looking west

Westbrook   41°N
(25 mi. E of New
Haven)

Building: 1½-story, 1900 sq. ft.
Collector: 450 sq. ft., water type
Storage: 2000 gals. water
% Solar-heated: 70 (estimated)

House 70% solar-heated by 450-sq.-ft. water-type collector and 2000-gal. storage system

Building   This three-bedroom house has a heated basement, a horizontal roof, and, at the northwest corner, a built-in, unheated 2-car garage. The house faces 10 degrees west of south. The walls and roof are well insulated. Windows occupy 60 percent and 70 percent of the south sides of the first and second stories respectively. All windows are double-glazed.

Collection   The 450-square-foot water-type collector consists of three arrays, sloping 57 degrees, mounted on the horizontal roof. The southmost array is 28 feet long and the other two are 24 feet long. The arrays are parallel and approximately 16 feet apart. Each is 6 feet high. Each faces 10 degrees west of south (as does the house also). The panels are of an early design by Sunworks, Incorporated. Each is 6 by 2 feet and employs, as absorber, a .017-inch copper sheet to which ¼-inch-diameter copper tubes, 5 inches apart on centers, have been soldered. The black coating is somewhat selective, having an a/e ratio of approximately 5. The glazing is single. The panel backing includes 3 inches of fiberglass. The coolant is water to which propylene glycol has been added as antifreeze, this material being much less toxic than ethylene glycol. Heat is delivered to the storage system via a heat exchanger.

Storage   The 2000-gallon storage tank is a steel cylinder 12 feet long and 5 feet 4 inches in diameter, oriented with its axis horizontal. It is in an insulated enclosure between the basement and the garage. Heat is distributed to the rooms by a fan-coil system.

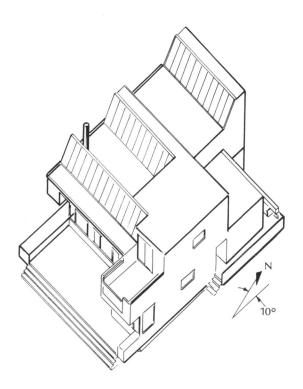

Auxiliary Heat Source   An oil furnace.

Domestic Hot Water   Not currently solar-heated, as explained below.

Cooling in Summer   Little is needed, especially as the house is within 50 feet of the ocean and the ground-floor windows are shaded in summer by a 4-foot roof overhang. There is no air conditioner.

Problems and Modifications   It was presumed necessary to add corrosion inhibitor to the coolant used, and accordingly sodium chromate was added. This chemical is toxic, and because there was thought to be risk that it might find its way into the domestic hot water, the practice of preheating the domestic hot water by means of the main solar heating system was discontinued. Late in 1977 a new scheme for solar preheating of this water was being planned.

Solar engineer: E. M. Barber, Jr. *Architect:* Donald Watson. *Owner and occupant:* Not stated. *Cost of house and solar heating system:* About $60,000. *Funding:* Private.

Atlanta   33½°N

Building: 1-story, 32,000 sq. ft.
Collector: 10,000 sq. ft., water type
Storage: 45,000 gals. water
% Solar-heated: 60

Near-million-dollar solar heating and cooling system for large school in Georgia

Building   This one-story, 32,000-square-foot building, with horizontal roof, was constructed in 1962. It is used the year around by about seven-hundred students. The winter heating load and summer cooling load are approximately equal. The solar heating and cooling system was formally accepted by the Atlanta Public School Administration on January 10, 1977.

Collection   A total of 576 collector panels, made by PPG Industries, Incorporated, are used. They are arranged in twelve rows, each sloping 45 degrees, mounted on the horizontal roof. Each panel is 76 by 34 inches. The absorber sheet is an Olin Brass Company aluminum Roll-Bond sheet with integral passages for coolant. The thickness of the passage walls is .030 inch and the thickness in regions between passages is .060 inch. The inside dimensions of a passage are .375 inch by .065 inch. An Alcoa selective black coating is used. The glazing is double, consisting of two sheets of 1/8-inch PPG Herculite tempered glass in standard aluminum frames. The coolant is water to which an inhibitor (capable of reducing the rate of corrosion to less than .0001 inch per year) has been added. No antifreeze is used; the coolant is drained at the end of the day and is replaced with nitrogen gas.

Reflectors.   Adjacent to each row of collector panels there is an oppositely sloping crude reflector which directs additional radiation to the collector and also helps shade the roof. Because the reflectors slope downward toward the north and are 36 degrees from the horizontal, these functions are most effective in summer, when the sun is high in the sky. This is appropriate, especially as the need for power to operate the cooling system is great in the summer. The reflector

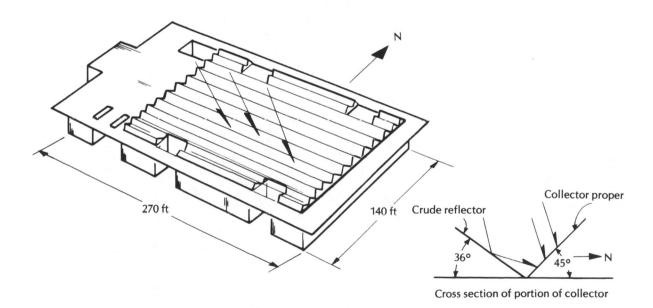

270 ft   140 ft   Crude reflector

Collector proper

36°   45°   N

Cross section of portion of collector

consists of a sheet of aluminized mylar bonded between clear mylar sheets and supported by a 1/8-inch sheet of Masonite. The total area of reflector is 12,000 square feet. The reflectance is about .74. The irradiation gain factor is 1.3 in midsummer and small (near unity) in midwinter. The typical collection temperature in midwinter is 140°F; in midsummer it is 200°F.

Storage   The 45,000 gallons of water is stored in three 15,000-gallon steel tanks, which are insulated and are located underground.

Auxiliary Heat Source   Existing gas furnace.

Domestic Hot Water   No information.

Cooling in Summer A 100-ton-capacity Arkla chiller is used. It employs lithium bromide and is powered by water at 195 to 200°F. Such water is supplied mainly by the solar heating system, but may alternatively be supplied by the above-mentioned auxiliary heat source. About 60 percent of the power needed for cooling is supplied by the solar heating system.

Problems and Modifications The mylar reflector used in the collection system has shown considerable deterioration and may need to be replaced after a few years of use. The practice of storing cold water in an insulated tank in summer, as part of the cooling system, has been discontinued.

Prime contractor: for equipment for the solar heating of the building, the solar heating of domestic hot water, and the solar powering of the cooling system: Westinghouse Special Systems, R. T. Duncan, Project Manager. *Solar architectural design:* P. R. Rittelmann of Burt, Hill, and Associates. *Mechanical design:* Dubin, Mindell, and Bloome, Associates. *Performance monitoring:* S. C. Bailey, et al., of Georgia Institute of Technology. *Total amount of funds supplied by NSF and ERDA:* $912,000. Note: of this, $180,000 was for design and $649,000 for initial construction.

Shenandoah Solar Recreation Center

$2,000,000 building, half underground, with $726,000 solar heating and cooling system employing 11,200-sq.-ft. water-type collector

Shenandoah   33°N
(32 mi S of Atlanta)
12032 Amlajack Boulevard

Building   This two-story, 54,000-square-foot building, used as a recreation center, includes office, lounge, meeting room, gymnasium, game room, and ice skating rink. Also there is a utility, or mechanical, room which contains may components of the solar heating and cooling system, including indicators and controls; this room constitutes a part of the lobby and may be viewed by visitors entering or leaving the building. The building is mainly of poured concrete, with sloping sides covered with earth. There is no basement. The window area is small, and all windows are single-glazed. They are screened from direct summer sun. The building has many special energy-saving features.

Collection   The 11,200-square-foot water-type collector is mounted on the sawtooth roof. The sixty-three panels, each 20 by 8 feet and tilted 45 degrees, are arranged in nine east-west rows of seven panels each. They were made by Revere Copper and Brass Company. The absorber is a .032-inch-thick sheet of copper with integral 3/8-inch-diameter copper tubes 5½ inches apart on centers. The selective black coating, of electroplated black chrome, has an a/e ratio of .95/.07. The panels are double-glazed with double-weight tempered low-iron glass. Polysulfide sealant is used. The coolant is water with

Building: 2-story, 54,000 sq. ft.
Collector: 11,200 sq. ft., water type
Storage: 15,000 gals. water
% Solar-heated: 95 (predicted)

no antifreeze; in cold sunless periods hot water from the storage tank is circulated intermittently and briefly through the panels to prevent freeze-up. Water flows through the panels in parallel. The maximum flowrate is 315 gpm. A 20-hp centrifugal pump is used. Hot water from the collector flows directly into the storage tank; no heat exchanger intervenes.

Reflector. Just south of each row of panels there is a row of polished aluminum reflectors, of ALCOA Coilzak, that slope downward toward the north at 36 degrees from the horizontal. The reflector area is 26,000 square feet and the reflectance is about 85 percent. The reflector is helpful mainly in the summer, when the sun is high in the sky. The amount of hot water needed for powering the cooling system in summer far exceeds the need for hot water for space heating in winter.

Storage The 15,000 gallons of hot water is stored in a pressurized horizontal cylindrical steel tank buried within a berm adjacent to the southeast corner of the building. The tank is insulated with 2 feet of dry sand. There is also a 2300-gallon buffer tank with 9 inches of insulation. The rooms are heated by fan-coil systems.

Auxiliary Heat Source A gas heater.

Domestic Hot Water This is heated in a coil immersed in the above-mentioned buffer tank.

Servicing the Ice Skating Rink Hot water from the solar heating system is used periodically to melt the uppermost portion of the ice in preparation for forming a new smooth surface.

Servicing the Swimming Pool In the spring and fall the pool is heated by the solar heating system.

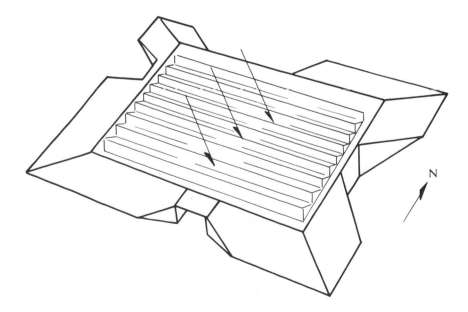

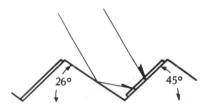

Cross section of a portion of the collector

Cooling in Summer   A Trane Model C-2A absorption-type cold generator, rated at 93.4 tons and 5.4 kw, is used. Normally it is powered by water at or above 190°F received directly from the collector. There are two tanks for storage of chilled water; each is a 30,000-gallon steel tank that may be pressurized to 25 psi. Also, the above-mentioned 2300-gallon tank is used in the primary loop to reduce the tendency of the chilling system to cycle on and off excessively. The unwanted heat is dissipated by a conventional cooling tower. The solar heating system is expected to provide 65 percent of the energy required to operate the cooling system. The balance is provided by a gas-fired boiler.

Problems and Modifications   No information.

Prime contractor for solar heating and cooling system design and installation: Georgia Institute of Technology; J. R. Williams, Project Director. *General coordination:* Shenandoah Development, Incorporated (esp. Ray Moore). *Architect:* Taylor and Collum (esp. Richard Taylor). *Engineering specifications:* Newcomb and Boyd, Engineers (esp. M. George, W. Shiver, and S. Bruning). *Others assisting in planning:* James Craig, T. Hartman, and S. P. Kezios of Georgia Institute of Technology; W. Cohen of Delta Corporation; also R. Wright. *General contractor:* Batson-Cook Company. *Owner and occupant:* Shenandoah Development, Incorporated. *Funding of building proper:* $2,000,000 from HUD. *Funding of design, construction, and trial operation of the solar heating and cooling system:* $726,000 from ERDA; E. Doering, Project Manager.

Carlyle   38½°N
(50 mi E of St. Louis)

Building: 2-story, 650 sq. ft.
Collector: ⎫
Storage: ⎬ Passive
% Solar-heated: See text

Weekend house with transparent portion of roof served by combination reflector and shutter

Building   This two-story weekend-vacation house has two rooms and a bathroom. There is 650 square feet of heated living space and a 290-square-foot fenced-in wooden deck at the south. The second story is a 24-foot-diameter half-dome—part of an aluminized steel silo dome, with cylindrically curved, corrugated steel siding below. The first-story walls and floor include 5 inches of fiberglass insulation. The second-story walls and ceiling have 2 inches of sprayed urethane foam insulation to which a fire-resistant paint has been applied. There is no basement or attic. The house faces exactly south.

Passive Solar Heating System   Solar radiation is received by a small (120-square-foot), 30-degree-sloping, transparent roof area adjacent to the half-dome. The glazing is single, consisting of two 12-by-4-foot corrugated sheets of 5-ounce Filon (fiberglass and polyester), laid end-to-end. Just above this transparent roof area is a 24-by-5-foot reflector, hinged along its long north edge. It consists of aluminized mylar backed by canvas, plywood, and insulation. When tilted steeply upward toward the south, it directs much solar radiation downward toward the transparent roof and toward the small storage system 6 feet below it. The angle of the reflector can be adjusted

Reflectors open

manually from week to week or even from hour to hour if the occupant so desires. When the reflector is lowered and rests on the transparent roof, it serves as an insulating shutter. In the south central part of the first story there are recessed sliding glass doors of Thermopane, and on each of the adjacent south-face areas there is a 4-foot-diameter hemispherical plastic window. On the second-story south wall there is a 5-foot-diameter hemispherical plastic window. Each such window is provided with shutters for insulation on winter nights and with canopies to exclude solar radiation in summer. A small amount of thermal storage is provided by two liquid-filled tanks situated 6 feet below the transparent roof. Each tank is rectangular and is 8 by 2 feet by 1 foot high and rests on the floor. Each is of anticorrosion-coated galvanized steel and contains 100 gallons of a solution of 80 parts of water and 20 parts of ethylene glycol. The horizontal top of tank is a sheet of transparent corrugated plastic; 1½ inches below this there is a sheet of corrugated metal that has a nonselective black coating. Below this, immersed in the liquid, there is a vertical steel honeycomb that helps carry heat downward into the lower part of the tank. The sides and bottom of the tank are insulated. Solar radiation that has been directed downward by the reflector strikes the tanks and warms them, and accordingly the tanks help keep the house warm at night. Cushions may be laid on the tops of the tanks to insulate them—and to convert them to seats.

Reflectors closed

Percent Solar-Heated This has not been evaluated. Nor has the effectiveness of the small storage system described above been evaluated. The designer's guess as to percent solar-heated is 50. It is to be noted that this is a weekend-vacation house and is not expected to be occupied for long periods in winter.

Auxiliary Heat Source One portable electric heater.

Domestic Hot Water This is not solar-heated.

Cooling in Summer There is no formal cooling system. Little cooling is needed, inasmuch as the building is insulated and the windows are shaded, as explained above.

Problems and Modifications The initial insulation of the house was found to be marginal, and in 1977 more insulation was added.

Solar engineer, designer, builder, owner: Michael Jantzen. *Cost of materials for entire house:* $9000. *Incremental cost of solar heating system:* $350. *Funding:* Private.

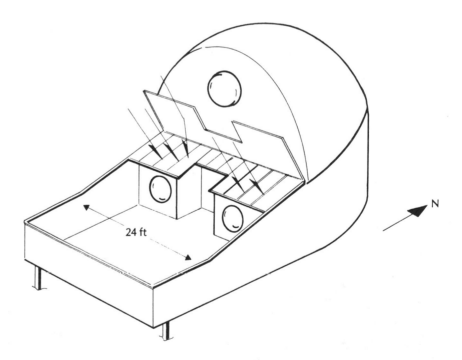

Eureka Solar House

House 60% solar heated by 800-sq.-ft. water-type collector and 1500-gal. storage system

Eureka   40°N
(120 mi SW of Chi-
cago)
Dennis Drive

Building   This house, a modification of a standard type, has a main story, attic, and walk-out basement. The living area, not including the basement, is 2000 square feet. The main story includes living room, kitchen, and two bedrooms. The attic includes a family room and two bedrooms. The walls and roof are insulated with 3½ inches of fiberglass and 7 inches of fiberglass, respectively. The building faces exactly south.

Collection   The collector, which is an integral part of the 45-degree roof, includes thirty-six panels, each of which is an Olin Brass Company aluminum Roll-Bond panel, 8 by 3 feet, with integral passages for coolant. The lower headers are of aluminum, but much of the piping in the collection system as a whole is of copper. The black coating used is nonselective. Each panel is double-glazed with tempered glass and has a 2-inch fiberglass backing. The coolant, equal quantities of water and ethylene glycol with a small amount of corrosion inhibitor, is circulated by a 1-hp centrifugal pump. Heat is delivered to the storage tank via a heat exchanger.

> **Building:** 1½-story, 2000 sq. ft.
> **Collector:** 800 sq. ft., water type
> **Storage:** 1500 gals. water
> **% Solar-heated:** Approximately 60

Storage   The 1500 gallons of coolant is contained in a fiberglass-insulated steel tank in the basement. The tank is 6 by 6 feet by 6 feet high. Heat is distributed to the rooms by a fan-coil system. The automatic control system, designed by Sun Systems, Incorporated, provides five modes of operation of the heating system as a whole.

Auxiliary Heat Source   Electric resistance heaters.

Domestic Hot Water   This is preheated by the solar heating system. A special heat exchanger is used.

Cooling in Summer   No information.

Problems and Modifications   After the system had been in operation for two years, some solid particles accumulated in the coolant. This coolant was then discarded and new coolant was introduced.

Solar engineer: Y. B. Safdari. *Architect:*   Tom Landes, Jr. *Builder and owner:*   Sun Systems, Incorporated.

2370-sq.-ft. house, with three studios, heated by combination of passive system and two separate air-type active systems

38½°N
South part of state

Building   This two-story, two-bedroom house has a gross area of 2370 square feet and a heated area of 2100 square feet. It faces exactly south. The north face of the house is vertical, but most of the other faces are curved and sloping. There is no attic or garage. There is a crawl space but no basement. The first story includes a living-kitchen-dining room, bathroom, and two studios—at the east and west ends. The second story includes a bedroom, studio, and, at the north side, a deck. In the south central part of the building there is a clerestory region 16 feet long and 6 feet wide. Most of the curved faces of the building derive their strength from laminated wooden arches. Most of the facing is of factory-painted corrugated galvanized steel, with the corrugations vertical. The footings and foundation walls, of poured concrete and concrete blocks, are insulated on the inside with 2 inches of Technifoam. Walls are insulated to R-21 with 3 inches of foam boards and a ½-inch airspace. Roofs are insulated to R-28 with 4½ inches of foam. The floor of the first story is insulated with 1 inch of foam and 8 inches of fiberglass. The total area of vertical windows is 213 square feet, of which 200 square feet is on the south side. The central window area on the south side of the first story is 16 feet long and 5 feet high; it is single-glazed with bubble-type domes of ¼-inch Plexiglas. Also, there are two simple circular windows which serve the studios at the ends of the house. The central window area of the second story is 8 by 8 feet and is single-glazed. At night, all large windows are covered by hinged insulating shutters of rigid foam.

Building:	2-story, 2370 sq. ft.
Collector:	Combined passive and active; see text
Storage:	
% Solar-heated:	See text

Collection   The solar heating system includes a passive system and an active system. Their contributions are in the ratio of 1 to 2 (predicted).

Passive solar heating system.   Solar radiation enters via a total of 200 square feet of south windows, of which 135 square feet is on the first story and 65 square feet is on the second story. The first-story south-window area consists mainly of the set of Plexiglas domes mentioned above. The radiation entering here strikes a 12-by-12-foot, 2-inch-thick pad of bricks, which contributes moderately to thermal storage. Radiation entering the second-story window area travels about 6 feet and strikes a balcony wall serving the clerestory; this wall has a ½-inch-thick stone covering which contributes slightly to thermal storage.

Active solar heating system.   There are two air-type collectors, fabricated on-site, with a combined area of 320 square feet. They are situated on the south face of the house, on either side of the central window area, where the exterior face slopes and curves. The east collector is 8 feet wide and 20 feet high; the slope is variable, a typical value being 50 degrees. The absorber is a sheet of corrugated galvanized steel that was painted at the factory with a nonselective black paint. The corrugations are vertical. The glazing consists of a single sheet of .040-inch Kalwall Premium Sun-Lite spaced 1 to 1½ inches from the black sheet. The panel backing includes 1 inch of high-density fiberglass and 3 inches of sprayed urethane foam, the latter being protected by a ½-inch layer of troweled Pyrocrete (a nonflammable material made by Carboline Company). A small blower circulates air upward in the 1-inch space between black absorbing sheet and the insulating backing. The emerging hot air may be directed to the storage system or to the rooms. The ducts used are 12 by 12 inches in cross section and are insulated with 1 inch of high-density fiberglass. The collector may also be operated in thermosiphon mode, to supply heat to the second-story rooms. The west collector is similar. The two collectors operate independently: one can run while the other is not running; one can serve its storage bin while the other is serving the

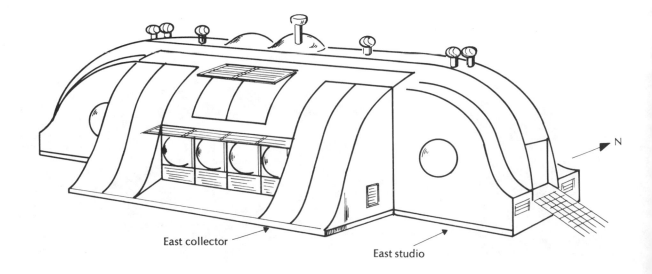

East collector

East studio

rooms directly. In summer each collector is vented by ports at top and bottom.

Storage   Two bins-of-stones are provided for the active system; they include, in all, 15 tons of 4-inch-diameter stones. One of the bins is beneath the southeast corner of the central living area and the other is beneath the southwest corner. Each is 11 by 9 feet by 1½ feet high Each is insulated with 4 inches of Technifoam. Hot air from each collector travels north through its respective bin. When the rooms need heat, room air is circulated through the bin in the same direction by a small blower.

Percent Solar-Heated   No estimate is available. The designer expects the value to be large.

Auxiliary Heat Source   A small, high-efficiency wood-burning stove. There is no furnace or electric heater.

Domestic Hot Water   This is preheated by a separate solar heating system made on-site. The collector, 64 square feet in area, is situated just outside the house at the base of the central first-story window area. A small pump circulates water from the collector to a 40-gallon tank. Final heating is by a wood-burning stove or by two electric in-line heaters.

Cooling in Summer   No formal cooling is provided. The south windows are protected by insulating shutters. Vents allow hot air in the upper parts of the house to escape. There are twelve turbine-type vents on the roof.

Problems and Modifications   Consideration has been given to adding a second glazing sheet to the large, single-glazed windows.

Solar engineer, designer, builder, owner, occupant: Michael Jantzen. *Cost of materials for the collectors:*   Of the order of $1 per square foot. *Funding:*   Private.

Indianapolis   40°N
6010 Southeastern
Avenue

Building: 1-story, 1530 sq. ft.
Collector: 378 sq. ft., air-type
Storage: 33 tons of stones in three bins
% Solar-heated: 80 (predicted)

System employing 33 tons of heat-storage stones in three bins each contiguous with two rooms, heating them directly by conduction and radiation

Building   This well-insulated brick building houses a dental clinic. There is an attic but no basement or garage.

Collection   The twenty-four Solaron-type collector panels, employing absorber plates of black aluminum, are mounted on the steeply sloping south roof.

Storage   The 33 tons of fist-size stones are in three vertical cylindrical bins situated near the center of the first story. They form a triangular array and are 6 to 12 feet apart. Each bin is contiguous with two rooms and a hallway, heating them directly by conduction and radiation. Nearly every room is so served. The bin temperatures are expected to reach 160 to 180°F under favorable circumstances.

Auxiliary Heat Source   A Bryant air-to-air heat-pump of 4-ton capacity. Supplementary electrical heating is incorporated in the heat-pump system.

Domestic Hot Water   This is not solar-heated.

Cooling in Summer   No information.

Problems and Modifications   The control system was defective, causing the rooms to become overheated. Repair was soon made. The insulation on the collector and the air ducts was judged to be inadequate and was improved. The thermostat and control system was modified so as to provide a half-hour delay after the solar heat distribution system was turned on and before the heat-pump was turned on.

Solar consultant: Jerome Kaplan. *Architect:*   Gordon Clark Associates. *Owner:*   J. T. Bohnert. *Supplier of collector:*   Solar Energy Products Company.

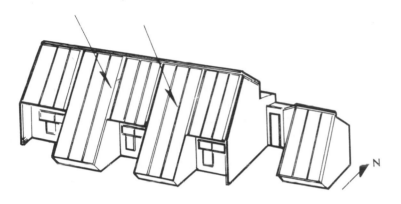

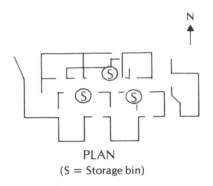

PLAN
(S = Storage bin)

Hayes Arboretum Solar Greenhouse

Richmond   40°
(70 mi E of India-
napolis)
801 Elks Road

Building: 1-story, 1000 sq. ft.
Collector: Spray type; see text
Storage: 5700 gals. water
% Solar-heated: 70-90

In this solar greenhouse the black absorbing surface is close to the north wall; that is, the useful space is between the glazing and the black absorbing surface. Heat is picked up from this surface by a spray of water.

Building   This is a steel-framed, 50-by-20-foot greenhouse, which faces exactly south. The south face, which slopes 53 degrees, is 50 feet long and 18 feet high. It consists of Beadwall. During the day, solar radiation passes through the two sheets of Tedlar-coated 5-ounce reinforced fiberglass and through the 2-inch space between them. At night this space is filled with 1/8-inch-diameter beads of polystyrene foam. The concrete-block foundation walls and the north wall are insulated with Zonolite and 1 or 2 inches of Styrofoam. There is no attic, basement, or garage.

Collection   Much of the radiation entering the building strikes a vertical .006-inch sheet of black nylon-reinforced polyethylene that is suspended 18 inches from the north wall. Water (with no antifreeze) is sprayed southward onto this sheet by nineteen nozzles affixed to the north wall. A ½-hp centrifugal pump is used, and the flowrate is variable. The water that trickles from the lower edge of the sheet is discharged into the storage tank. Some solar radiation strikes the growing plants, and some strikes the floor, which consists of a 4-inch slab of poured concrete resting on 3 feet of gravel.

Storage   The 5700 gallons (48,000 pounds) of water is in a rectangular poured-concrete tank, 48 by 4 feet by 4 feet high, situated beneath the absorber sheet. The tank walls are 12 inches thick. The north side of the tank is insulated with 2 inches of Styrofoam, and the bottom is similarly insulated. On the south side there is no insulation; heat can escape into the concrete floor and into the gravel beneath the floor. The safety cover of the tank is made of wooden boards and has no added insulation. Heat from the tank is distributed to the envi-

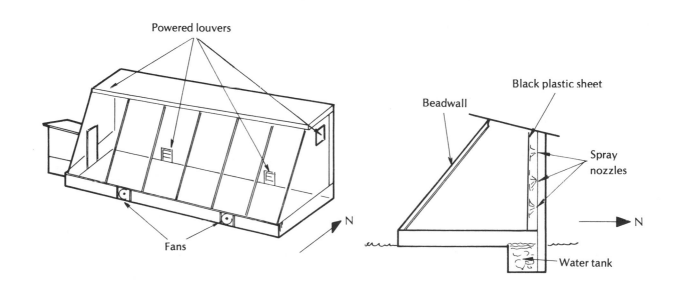

Powered louvers

Fans

Beadwall

Black plastic sheet

Spray nozzles

Water tank

rons of the potting benches by 200 feet of finned tube running under the benches. The water is circulated through the tube by a small centrifugal pump. A SERC Thermo-Mate Controller DC-761 controls all of the main components of the collection and distribution system.

Auxiliary Heat Source   A natural-gas water boiler.

Domestic Hot Water   None.

Cooling in Summer   Alternate pairs of Beadwall panels are kept filled with beads at all times to provide partial shade.

Problems and Modifications   Initially the Beadwall panels failed to perform properly. The difficulty was soon corrected.

Solar designer and general supervisor: D. R. Hendricks. *Builder:*   C. Beard (masonry); arboretum staff (other). *Architect:*   Terwilliger Architects (esp. D. Sweet). *Cost not including labor:* $16,000. *Cost including labor:* $25,000. *Funding:*   Hayes Research Foundation.

Caivano House

Bar Harbor   44½°N
(110 mi NE of Port-
land)

Building: 2-story, 1300 sq. ft.
Collector: }
Storage: } See text
% Solar-heated: 60 (predicted)

House employing a balanced mix of low-cost solar heating systems, including direct passive, solar room [greenhouse], and Trombe-wall quasi-passive, with much long-term energy storage beneath concrete floor

Building   This 32-by-24-foot, two-bedroom house has a small attic but no basement or garage. The roof is supported by homemade, 10-inch-deep, laminated wooden arches resting on projecting 2-by-10-inch floor joists and is insulated with 10 inches of fiberglass. The east, north, and west walls are made of concrete blocks, and the outer faces of these walls are insulated with 2½ inches of sprayed-on ure-thane foam protected on the outside with 1/8 inch of Blockbond, a commercial stucco that contains fiberglass. The first story includes a large kitchen-dining-living room and also a two-story-high green-house 20 feet long and 4 feet wide. It is single-glazed with 1/8-inch Plexiglas. Between the greenhouse and the main room there is a massive wall-and-chimney, discussed below. The view windows are dou-ble-glazed: they are of Thermopane. The first-story windows are cov-ered at night with quilted dacron curtains. The first-story floor is a 6-inch slab of poured concrete, and it rests on a 2-foot-thick bed of gravel that is 30 feet long and 22 feet wide.

Solar Heating System   This system includes direct passive heating, heating of the greenhouse, and a quasi-passive system employing a massive wall and a massive floor.

Direct passive solar heating.   Much radiation enters the first-story and second-story windows at the east end of the south face of the building and heats the rooms there directly.

Solar heating of greenhouse.   Much radiation enters the greenhouse via its single-glazed south face, 20 feet long and 12 feet high. A frac-tion of this radiation is absorbed by the plants and the earth.

Quasi-passive solar heating.   Most of the radiation that enters the greenhouse passes through its 4-foot airspace, passes through an-other 1/8-inch sheet of Plexiglas, and—four inches beyond—strikes a massive wall-and-chimney. This structure, 12 feet long, 8 inches thick, and 14 feet high, is made of gravel-filled concrete blocks and is painted brown on the south side. The structure is warmed by the solar radiation absorbed, and it warms the air in the 4-inch space immedi-ately south of it. This warm air is forced downward, by a small blower, then northward within many pipes embedded in the concrete floor. These pipes, which are 6 inches in diameter and are of galvan-ized steel, fan out so as to serve nearly all portions of the massive floor and underlying gravel bed. The total length of pipe used is 200 feet. The air emerging from the pipes enters the room and flows back informally, via the room, into the greenhouse. The massive wall-and-chimney structure receives energy not only from the incident solar radiation but also from the wood-burning stove immediately to the north of it. The blower, which serves the sole purpose of transferring heat from the wall-and-chimney structure to the massive floor and gravel bed, is turned on automatically whenever that structure is hot-

ter than 80°F; thus it normally runs only when the sun is shining and electricity use is off-peak.

Auxiliary Heat Source   Wood-burning, cast-iron Lange stove.

Domestic Hot Water   This is not solar-heated.

Cooling in Summer   None is provided. Little cooling is needed, inasmuch as (1) the location—on the Maine coast—is a cool one, (2) the house is well insulated, (3) the roof overhang shields the living area from direct radiation in summer, and (4) the massive floor and also the massive, externally insulated walls have large thermal capacity.

Designer, owner, occupant: Roc Caivano. *Solar consultant:*   R. C. Hill. *Other consultant:*   H. Beal. *Builder:*   J. Krajewski. *Cost of house and solar heating system, exclusive of design work and some free labor:* $28,000. *Funding:*   Private.

East Raymond   44°N
(25 mi   NW of Port-
land)
On Tarklin Hill

Building: 2-story, 2400 sq. ft.
Collector: ⎱
Storage: ⎰ See text
% Solar-heated: 75-80 (predicted)

Combination of passive and active systems provides 75 to 80 % of the winter's heat need

Building   This three-bedroom house, of wood, stone, and concrete, is recessed into the south slope of a hill. The living room has a clerestory. The southwest part of the house includes an attached 15-by-10-foot greenhouse, which helps heat an adjacent woodworking shop. The house has a loft and a small attic space. A two-car garage is attached to the west end of the building. There is no basement.

The first-story floor consists of ¾-inch-thick slate resting on a 6-inch concrete slab, which in turn rests on a 12-inch bed of gravel. The edges of the slab are insulated with 2 inches of rigid foam. Some of the exterior walls are of 12-inch-thick concrete and are insulated on the outside with 2 inches of urethane foam. Other exterior walls are made with 2 x 8's and are insulated to R-24 with 7½ inches of foil-faced fiberglass. The roof, made with 2 x 12's, is insulated to R-38 with 11 inches of fiberglass batts. A polyethylene vapor barrier is used on all exterior walls and on the roof.

The area of vertical glass windows and glass doors on the south side of the main structure is 300 square feet. The area includes three sliding-glass-door areas, each 9 feet wide and 8 feet high, provided with especially positive seals. In includes also a row of small windows, situated above the active collector, which serve the living-room clerestory. All windows and glass doors are double-glazed. The three sliding-glass-door areas are recessed between vertical concrete columns, or fins, 2 feet in north-south dimension, which project 1½ feet to the south and serve to protect the glass doors from wind and help exclude solar radiation in summer. The combined area of the windows on the east, north, and west side of the building is small.

Collection   This includes an active system and a passive system.

Active solar heating system.   The gross area of the water-type collector is 600 square feet and the net area is 500 square feet. The collector is 38 feet wide and 16 feet high and forms an integral part of the roof, which slopes 60 degrees. It was constructed on-site. Thirty-eight Kennecott Copper Corporation absorber plates are used. Each is 22 inches wide and 96 or 76 inches long and consists of a .003-inch-thick copper sheet to which rectangular-cross-section copper tubes, 2 1/8 inches apart on centers, have been soldered. The black coating is nonselective. The glazing is double; the outer sheet is .040-inch Kalwall Sun-Lite (polyester and fiberglass) and the inner sheet, spaced ½ inch from it, is of Teflon. The tubes are hydraulically in parallel. The coolant is water, with no antifreeze; the system is drained before freeze-up can occur. The water is circulated at 20 gpm by a 1/6-hp centrifugal pump. Hot water from the collector flows directly into the storage tank; there is no heat exchanger.

Passive solar heating system.   This makes use of the above-described 300 square feet of living-area south glass doors and south windows, and also the greenhouse window area.

Storage   The storage system for the active solar heating system consists of a water-filled, rectangular 5000-gallon tank situated on the ground floor at the northwest corner of the main structure. It is of concrete, cast on-site, and is insulated with 2 to 4 inches of urethane foam. Heat from the tank is distributed to the building in two ways: (1) via a serpentine-pattern copper tube within the first-story floor slab, and (2) via a fan-coil system serving ducts that extend beneath the edges of that slab. The return air makes use of a vertical duct incorporated in the central fireplace-chimney structure discussed below; hot air in the upper part of the clerestory is drawn into this duct and conveyed downward (and picks up heat from this structure, if there is a fire in the fireplace). Storage is provided for passively collected energy by the above-described massive floor and massive exterior walls and also by the central fireplace-and-chimney structure, which has a mass of 10 tons. The total mass contributing to such storage is 120 tons.

Auxiliary Heat Source   A fireplace which has glass doors and a specially ducted supply of outdoor air. Also a wood-burning kitchen stove. Also a 15-kw electrical heater associated with the above-mentioned fan-coil system.

Domestic Hot Water This is preheated by the solar heating system: the water passes through a small preheat tank immersed in the main storage tank. Final heating is electrical.

Cooling in Summer None is needed and none is provided. Overhang shades the main south windows. Insulation helps exclude heat. The thermal mass of the house keeps the temperature rise small. Hot air is vented via clerestory louvers just above the main collector.

Solar designer and architect: Solar Design Associates (esp. S. J. Strong). *Builder:* L. B. Whitney and Company. *Owner and occupant:* R. L. Whitney. *Funding:* Private.

Maine Audubon Building

Falmouth Foreside 44° N
(6 mi N of Portland)
118 Old Route One,
at Gilsland Farm, on
north bank of
Presumscot River
Estuary near Casco
Bay

Building: 2-story, 5500 sq. ft.
Collector: 2000 sq. ft., air type
Storage: 105-ton bin-of-stones
% Solar-heated: 75

Large headquarters building with 2000-sq.-ft. air-type collector fabricated on-site [at $4.15 per sq. ft.] and 105-ton bin-of-stones

Building   The wood-frame structure, with an overall length of 80 feet and an overall width of 64 feet, bears some resemblance to a New England barn and to a saltbox house. At the west end is a large cutout area where a slender, 26-foot-long extension at the southwest corner houses a wood-burning furnace and provides a windbreak for the entryway. All living space is contained in the main part of the building, which has a concrete floor slab insulated along its east, north and west sides with 2 inches of urethane foam. The south side contains the storage system. Walls are insulated with 5½ inches of fiberglass and the roof with 9 inches. Incorporated in each exterior wall, behind the sheetrock facing, is a continuous sheet of polyethylene that prevents ingress of air and moisture. The building faces directly south, and window area on the east, north and west sides is small, with a combined area of 370 sq. ft. These windows are triple-glazed. On the south roof of the building, about 30 feet from the east end, is a band of panels 16 feet wide and 30 feet high overall. The central portion, 8 feet wide, is translucent. It is glazed with five sheets of Kalwall Sun-Lite. The two flanking portions of the band are transparent and are triple-glazed with glass. The entrance door is provided with air lock. The goal in controlling leakage in the house was to reduce the rate of exchange of air to one exchange per hour. The goal in controlling conductive heat loss was to reduce this loss to 40 percent of the total loss.

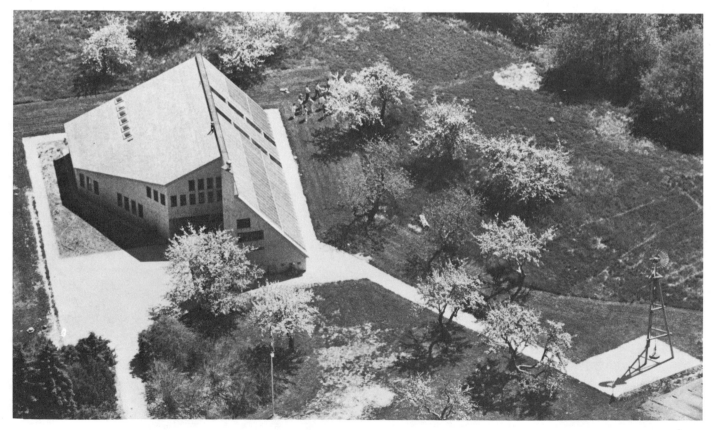

Collection The collector is located on the south roof, with a slope of 50 degrees, and measures 80 by 34 feet including the gap occupied by the light-transmitting panels described above. The net area is 2000 square feet. The collector proper contains thirty-four panels, each 30 by 2 feet, which were built on-site at a cost of $4.15 per square foot. The heart of each panel is a sheet of ½-inch CDX plywood covered with a coat of nonselective black paint. Stapled to the upper face of the plywood, between the upward-projecting 2x4's that define the panel, is ordinary aluminum mesh window screen that has been formed into a fairly tightly pleated surface and painted black. This wavy screen increases the area of absorption of radiation, increases the area of heat transfer to the air, and increases the turbulence of the airflow in the collector. However, it also increases pneumatic resistance. Glazing is a single sheet of .040-inch Kalwall Premium Sun-Lite 3½ inches above the plywood sheet. Air flows upward within the panel, then returns in a space immediately below the plywood sheet. Affixed to the underside of the collector assembly is a 9-inch layer of fiberglass with a foil face on its underside. Hot air from the base of the collector assembly flows north into a tan chamber, whence it is blown downward into the bin-of-stones by three 1-hp fans, spaced uniformly along the 80-foot-long region below the collector, at a combined flowrate of 11,000 cfm. In addition, some solar energy enters the south central portion of the building's living space directly through the band of light-transmitting panels described above.

Storage A rectangular bin of poured concrete, 78 by 9 feet by 10 feet high, extends the full length of the south side of the building, with its floor 4 feet lower than that of the living space. It is filled to a depth of 3 feet by pieces of crushed rock 1 to 1½ inches in size, beneath which is an 8-inch-high plenum defined by a mesh of steel bars and screen. The upper part of the bin is largely empty, serving as a plenum and providing space for delivery of heat from the auxiliary furnace. The exterior walls of the bin are insulated with 4 inches of urethane foam. Fan-driven hot air from the collector flows downward into the bin and through the stones, returning via a distribution plenum on the lower edge of the collector, through which it again flows as described above. The calculated pressure drop in the air traveling

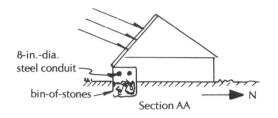

8-in.-dia. steel conduit
bin-of-stones
Section AA
N

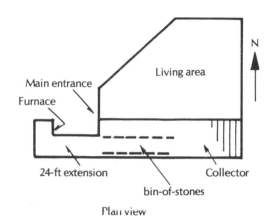

Main entrance
Furnace
Living area
24-ft extension
bin-of-stones
Collector
N
Plan view

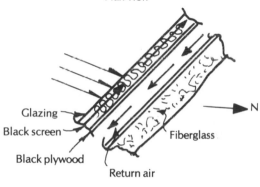

Glazing
Black screen
Black plywood
Fiberglass
Return air
N

Cross section (looking west) of portion of collector

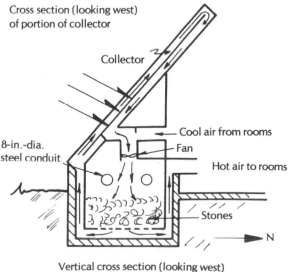

Collector
Cool air from rooms
8-in.-dia. steel conduit
Fan
Hot air to rooms
Stones
N

Vertical cross section (looking west) of collector and bin-of-stones

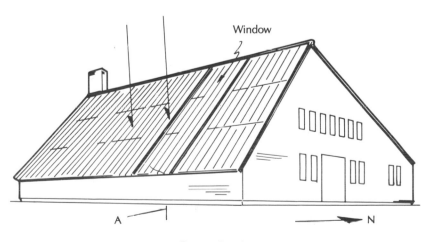

Window
A
N

Perspective view

downward through the stones at a nominal velocity of 15 feet per minute is 0.1 inch of water. When rooms need heat, a special 1-hp blower circulates room air downward through the bin-of-stones at 2000 cfm and thence back into the rooms. Alternatively, the somewhat hotter air just above the stones can be circulated by the blower, to the rooms by operation of appropriate dampers, which is done automatically whenever the temperature at the base of the bin falls below a specified value.

Auxiliary Heat Source   A specially designed wood-burning furnace is located in the extension at the southwest corner of the building, outside the living space. Designed for high-efficiency combustion, the furnace has an insulating liner of refractory brick, so that very little heat escapes except with the exit gases and combustion proceeds at such a high temperature that almost no creosote, tar, or smoke is emitted from the chimney. The large firebox can be loaded with 200 pounds of wood at one time, and this quantity can be consumed in one hour. During a typical winter, four cords of wood are burned. Heat from the furnace is delivered to the bin-of-stones via a loop of 8-inch galvanized steel conduit (culvert pipe) 150 feet in length, in the shape of a long, narrow U, mounted horizontally in the bin in the airspace just above the surface of the stones. The hot conduit warms the enveloping, downward-flowing air, which in turn warms the stones below. Alternatively, the hot air above the stones may be delivered directly to the rooms as explained above. Final discharge of furnace exit gases, which may be cool by the time they enter the vertical discharge stack, is assisted by a blower.

Domestic Hot Water   This is preheated in a copper coil buried in the stones in the bin.

Cooling in Summer   Cooling is by natural ventilation, with the living-space venting taking advantage of the prevailing west wind. The collector is vented by large openings at top and bottom. No attempt is made to cool the bin-of-stones (by circulating cool night air through it) lest the cool stones condense moisture out of subsequent warm air circulated through them, thus encouraging the growth of fungi and bacteria in the storage system.

Problems and Modifications   A few small difficulties have arisen, most of which have been overcome. At first there was a moderate amount of leakage of hot air from the lower edge of the collector as a result of loosening of battens by snow and ice. Much of the air traveling upward in the collector panels failed to reach the upper end because it leaked through cracks between the black plywood sheets and into the return channels. There were some air leaks in the enclosure of the bin-of-stones. The thickness of insulation on the exterior walls of the bin is only marginally adequate.

Solar consultant: R. C. Hill. Architect: G. B. Terrien. Project supervisors for Maine Audubon Society: W. J. Ginn, E. Morgan. Designer and builder of furnace: R. C. Hill. Cost of building and solar heating system: $270,000. Cost of solar heating system alone, not including ducts, blowers and furnace: $25,300

Akehurst House

Baltimore   39°N
4138 Joppa Road

Building: 1½-story, 2700 sq. ft.
Collector: 512 sq. ft., trickling-water type
Storage: 1800 gals. water
% Solar-heated: 65

Retrofit system employing trickling-water-type collector, an 1800-gal. tank of water, and no stones

Building   The four-bedroom wood-frame house was built in 1949, and solar heating was installed in 1976. The first-story and second-story floor areas are 1900 and 800 square feet, respectively. There is a full basement, and no garage. The windows are supplemented in winter with storm windows. The window area is modest. The walls are insulated with 3½ inches of rockwool and the ceiling or roof with 4 inches of rockwool. Only the first story is heated directly by the solar heating system; the second story is served by it indirectly and only to a moderate extent. The building faces exactly south.

Collection   The Thomason-type trickling-water collector is mounted on the roof, which slopes 45 degrees. The heart of the collector is a corrugated sheet of .019-inch aluminum with valleys 2 5/8 inches apart on centers. The black coating is nonselective. Water is fed to the upper edge of the collector by a ¾-inch-diameter main pipe and several 3/8-inch-diameter branch pipes, which have one 1/16-inch-diameter outlet hole per valley. A gutter at the lower edge of the roof collects the water, which then flows into the storage tank. The water is supplied to the collector at 15 to 20 gpm by a ⅓-hp centrifugal pump. All of the pipes are of copper. The panels are single-glazed with single-strength glass sheets 47 by 24 inches. No antifreeze is used.

Storage   The 1800 gallons of water is contained in a 9-by-5½-by-5-foot concrete-block tank situated in the southeast corner of the basement. The walls of the tank are insulated on the inside with 2 inches of Styrofoam. A polyethylene liner is used. (No stones are used in

conjunction with this tank.) When the first-story rooms need heat, water from the storage tank is circulated by a 1/10-hp pump to the radiators in these rooms.

Percent Solar-Heated   Although the solar heating system supplies heat directly to the first story only, some heat flows from the first story to the second story. It is estimated that 65 percent of the winter's heat need of the house as a whole is supplied by the solar heating system.

Auxiliary Heat Source   Heat for the first story is supplied by an oil-fired boiler. Heat for the second story is supplied by the oil-fired boiler (mentioned below) that serves the domestic hot water system.

Domestic Hot Water   This is preheated in a 42-gallon steel tank within the main storage tank. Final heating is by a small oil-fired boiler of 30-gallon capacity.

Cooling in Summer   None.

Designer, builder, owner, occupant: P. F. Akehurst, of Akehurst and Sun. *Funding:* Private.

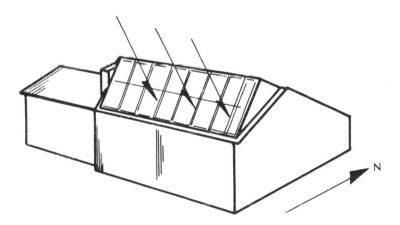

Calvert County 39°N
(15 mi SE of Wash-
ington, D.C.)
On Lyons Creek Road
in Dunkirk; off Route
4; 10 miles south of
Upper Marlboro.

Building: 2-story, 70 ft. long
Collector: 960 sq. ft., trickling-water type
Storage: 1600 gals. water and 28 tons of
stones
% Solar-heated: About 90-95 (predicted)

An excellent example of H. E. Thomason's versatile, simple, low-cost solar heating system employing a trickling-water-type collector

Building This two-story wood-frame house has overall length and width of 70 feet and 44 feet respectively. To define the heated area is difficult because of the different extents to which various parts of the house are heated. The first story, with a heated area of 1550 square feet, includes a living room, kitchen, dining area, four bedrooms, and two bathrooms. Also it has a solarium, solarium-pool room, and a two-car, tandem, drive-through, front-to-back garage. The south portion of the first story has a near-horizontal roof 70 feet long and 16 feet wide, which provides two sundeck areas. If, on a sunny spring day, the pool water is too cold, it is pumped onto the west sun deck, where solar radiation warms it, and is then returned to the pool. The east sun deck includes a hatch which, when open, reveals an access staircase and a chute (children's slide) to the pool. The second story, which is heated only by such heat as leaks upward from the first story or northward from the collector, includes a bathroom and has space that could be used for four bedrooms. The second-story roof is gabled at both ends and includes, on the north side, a transverse gable and two small dormer windows. The roofs slope 45 degrees. The basement, 54 by 26 feet, has an area of 1400 square feet, most of which is heated. It includes a 45-by-26-foot general purpose room that is heated by warm air and by the warm ceiling. Also it includes a 20-by-8-foot heat-storage system and a 56,000-BTU/hr oil-fired heater for heating domestic hot water and for backup space heating.

Insulation. The walls are insulated with 3½ inches of fiberglass and the ceiling is insulated with 6 inches of fiberglass. Most of the first-

story windows, and some other windows, are double-glazed. Most of the windows of the second story, basement, solarium, and garage are single-glazed.

Orientation. The building faces 15 degrees west of south, with the desirable consequence that the six-hour period during which the radiant flux per unit area of the collector is high is approximately synchronized with the period when the outdoor temperature is high, favoring high collection efficiency.

Collection The 60-by-16-foot trickling-water-type collector, on the roof sloping 45 degrees, has a gross area of 960 square feet and a net area of 800 square feet. It consists of fifteen panels, each 16 by 4 feet. The absorber is a corrugated aluminum sheet .017 inch thick, with the corrugations running up and down. The valleys, which are about ⅓ inch deep and 1 inch wide, are 2.7 inches apart on centers. The black coating is nonselective (or slightly selective). The collector is single-glazed with glass panes 48 by 24 inches. In the lowest 4-foot-wide band, where snow loads may sometimes be large, double-strength glass is used. In the upper part of the collector, single-strength glass is used. The panes are mounted in aluminum frames (with silicone bed gaskets) which rest on the corrugated aluminum sheet. The airspace between panes and aluminum sheet is about ¼ inch. The weight of a panel, including the glazing, is 2 pounds per square foot. The coolant, which is filtered rainwater with no antifreeze or inhibitor, is fed to the upper edge of the collector via a horizontal, ¾-inch-diameter copper distribution pipe and a set of junior pipes—one pipe per panel. Each of these pipes, also of copper, is 3/8 inch in diameter and has one 1/16-inch-diameter hole for each valley of the corrugations. Circulation is maintained by a ⅓-hp centrifugal pump that produces a flowrate of about 20 gpm. The pump runs whenever the temperature of the collector exceeds the temperature of the water in the bottom of the storage tank. The water that flows from the valleys is collected by a set of gutters (one gutter per panel) and is delivered to the top of the storage tank. The average temperature of the water flowing from the collector at noon on a sunny day in January is 80° to 110°F. The corresponding temperature in March is 90° to 125°F.

Storage The heart of the storage system is a horizontal, cylindrical, 1600-gallon steel tank containing water. The tank is 4 feet in diameter and 17 feet long and has a 24-inch-diameter manhole at top center. The tank is within, and near the bottom of, a bin that is 20 by 8 feet by 7 feet high and is situated in the west end of the basement. The bin has cinder-block walls and bottom, and it rests on a 5-inch-thick concrete slab, which in turn rests on a 1-inch layer of Styrofoam. The tank is surrounded by 28 tons of 1-to-2-inch-diameter stones. Above the stones is a crawl space and in the upper part of the space there is an array of finned copper tubes that serves as a part of the auxiliary heating system discussed below. The tubes are ¾ inch in diameter and the total length of tubing is 200 feet. Hot water from the collector enters the tank near the top, and water returned to the collector is taken from near the bottom.

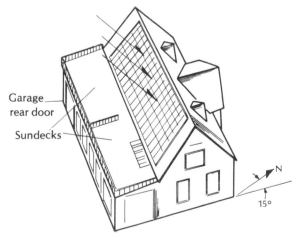

Garage
rear door

Sundecks

N

15°

Distribution. When the rooms need heat, a ¼-hp blower near the base of the bin draws cold room air downward (via a vertical duct 30 by 28 inches in cross section) and drives it horizontally into the base of the bin and thence upward through the quantity of hot stones. The air, now warm, flows horizontally throughout the 2½-inch space between the basement ceiling and the first-story floor, warming that floor. In mild weather the blower is not needed: adequate flow is produced by gravity convection alone. The ceiling mentioned is of aluminum-coated gypsum board.

Percent Solar-Heated   No figure is available. During an eighteen-month period of trial operation and testing by engineers from George Washington University, operation and maintenance were abnormal. A similar house, situated a few miles away and occupied, operated, and maintained by H. E. Thomason, is about 90 to 95 percent solar-heated.

Auxiliary Heat Source   There is no furnace. But heat can be supplied to the bin-of-stones by the oil-fired domestic hot water heater discussed below. The heat is distributed via the above-mentioned array of finned copper tubes; distribution starts whenever the pertinent solenoid-operated valve is opened. Also, in winter, some heat is recovered from the flue pipe of the oil-fired domestic hot water heater; the heater has two alternative flue pipes—for winter and summer—and the winter flue pipe passes through the bin and lies close above

Water trickling down
Sun's rays
Glass
¼ in
2.7 in
Cross section of small portion of collector

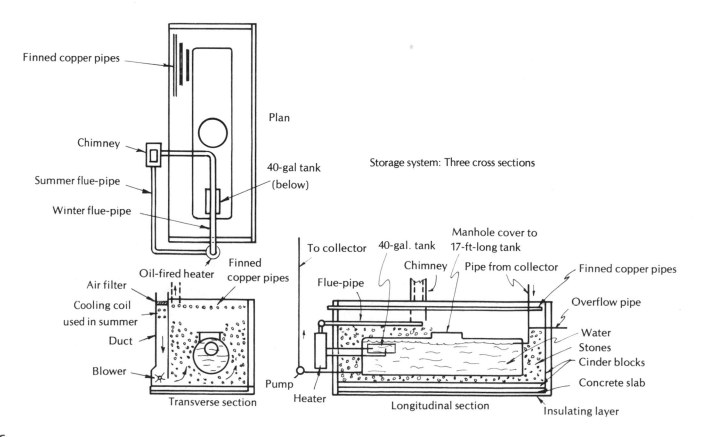

Finned copper pipes

Chimney

Summer flue-pipe

Winter flue-pipe

Plan

40-gal. tank (below)

Storage system: Three cross sections

Air filter
Cooling coil used in summer
Duct
Blower

Oil-fired heater
Finned copper pipes

Transverse section

To collector
40-gal. tank
Manhole cover to 17-ft-long tank
Chimney
Pipe from collector
Finned copper pipes
Flue-pipe
Overflow pipe
Water
Stones
Cinder blocks
Concrete slab
Pump
Heater
Longitudinal section
Insulating layer

the stones; this flue pipe is 8 inches in diameter and is of heavy (16-gauge) galvanized steel. The summer flue pipe, invoked by a manually operated damper, is 4 inches in diameter and is wholly outside the bin.

Domestic Hot Water   This is preheated in an uninsulated 40-gallon steel tank situated in the upper part of the main tank. Final heating is by a 56,000-BTU/hr oil-fired heater which has its own 40-gallon tank.

Cooling in Summer   In summer, a standard 28,000-BTU, $400, air conditioner is operated throughout the coolest hours of the night (e.g., off-peak hours from 10:00 p.m. until 4:00 a.m.) to dry and cool air that is circulated through the bin-of-stones. During a subsequent hot day, room air is circulated through the bin and is dehumidified and cooled thereby.

Humidification in Winter   In winter, room air circulated through the bin-of-stones is humidified by a controlled spray of water onto the stones. The operation is controlled by a humidistat and a timer.

Collection of Rainwater   Rainwater can be collected from the sundecks, the south portion of the main roof, and half of the north roof. The gutters for collecting rainwater are separate from the gutters serving the solar collector. Rainwater may be used to fill the main storage tank. If, in summer, the water in that tank is used to water lawns, the tank may be refilled next time there is a heavy rain.

Problems and Modifications   No simple account of operating problems and maintenance work can be given, because of complications associated with the performance studies made by engineers from George Washington University. In the similar house mentioned above (Thomason Solar House #5) no significant problems arose. One control device was found to be defective and was replaced free by the manufacturer.

Solar engineer, inventor, designer, and owner of many pertinent patents: H. E. Thomason. *Builder:*   Thomason Solar Homes, Incorporated. *Funding of performance studies made by George Washington University:*   By ERDA. *Cost of building and solar heating system [but not the land]:*   Approximately $80,000. *Cost of equipment for solar-heating the building, solar-preheating the domestic hot water, and cooling the building in summer:*   Approximately $4800. *Cost [in 1975] of collector panels:* $3 per square foot.

Crosley House

Royal Oak   39°N
(on Eastern Shore)

Building:	2-story, 1300 sq. ft.
Collector:	} Passive
Storage:	}
% Solar-heated:	50-60

House with two kinds of passive solar heating systems

Building   This two-bedroom wood-frame house has a small attic space and no basement. There is a two-car garage 20 feet to the north. The ground floor includes a living room, dining room, kitchen, bedroom, large south-facing bathroom, utility room, and (at the west end) a screened porch. The second story (at the east end of the house only) includes a combination bedroom-studio and a half-bathroom. The walls are insulated with 3½ inches of fiberglass and the roof includes 6 inches of fiberglass. The windows on the east, north, and west walls are small; their combined area, including 50 square feet of glass doors, is 155 square feet. The south windows are discussed below. All windows are double-glazed.

Solar Heating System   Two very different passive systems are used: a system employing a massive floor and a system employing a massive wall.

System employing massive floor.   This system extends along two thirds of the south portion of the house; it is in the southeast part of the house. The massive floor is 34 feet in east-west dimension and 14 feet in north-south dimension. Its thickness is 4 inches, made up of a 2-inch-thick upper face of brick resting on a 2-inch-thick slab of concrete. Each square foot of this floor weighs 50 pounds, and the total mass of the floor is 23,000 pounds. Beneath this floor there is a sheet of plywood and a 3½-inch layer of fiberglass. The floor is supported (and held off the ground) by joists. The massive floor receives and absorbs much solar radiation. Subsequently (mainly in the afternoon and evening) it gives out much heat to the rooms. The radiation is re-

ceived via vertical south windows. The full set of south windows has a gross area of 400 square feet and a net area of 350 square feet. Approximately half of this area (150 square feet, net) serves the massive floor. The windows are double-glazed with 3/16-inch tempered glass. The glass sheets are patio-door size, either 76 by 46 inches or 76 by 33 inches. The space between sheets is 2½ inches. On cold nights, heavy floor-to-ceiling curtains are drawn, to reduce heat loss.

System employing massive wall. This system, in the southwest part of the house (living room area) extends along one third of the south face. The vertical massive wall is parallel to the south windows and 3½ inches from them. The windows themselves are similar to those described in the previous paragraph. The wall is made of brown bricks and is 18 feet long, 8 inches thick, and 12 feet high. Its total mass is 16,600 pounds. It rests on concrete blocks supported by a poured-concrete footing. In the wall there are several openings (ports) which permit air to circulate, by gravity convection, from the living room into the 3½ inch space between wall and window, and vice versa. The ports are situated at various heights ranging from 6 to 9 feet above floor level. (They are so high up that they do not provide the occupants with a view of the landscape to the south.) Their aggregate area is 8 square feet. They remain open permanently, winter and summer.

Auxiliary Heat Source A wood-burning Franklin stove. Also electric resistance heaters incorporated in the gypsum-board ceilings.

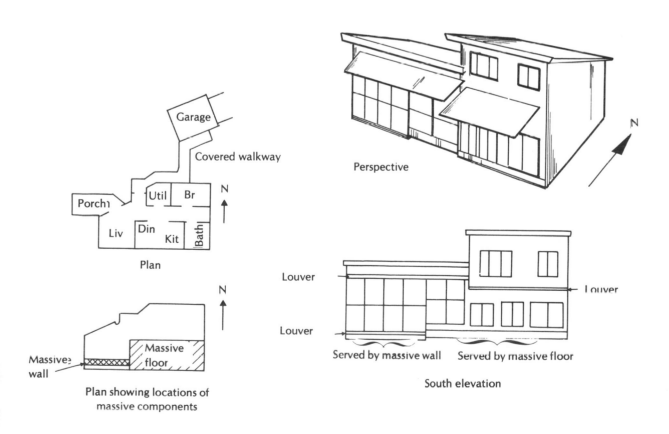

Perspective

Plan

Plan showing locations of massive components

South elevation

Domestic Hot Water This is not solar-heated.

Cooling in Summer A 32-inch-diameter attic exhaust fan is provided. In one bedroom there is a small conventional air conditioner. An extensive venting system is provided for the window area in front of the massive wall: there is a 14-foot-long, 6-inch-high vent (louver) above the top of the window area and a similar vent at the base of that area. A similar vent is provided along the top (but not the bottom) of the window serving the massive floor. Two large crank-down awnings shade all of the first-story south windows.

Problems and Modifications On winter nights some warm room air circulates via the ports in the massive wall into the space between wall and window, with some loss of heat. On some occasions moisture accumulates in the space between the inner and outer glazing sheets. Sometimes the north portion of the first story is slightly cold even when the south portion is amply warm. Methods of overcoming some of these difficulties are under consideration.

Solar designer: A. M. Shapiro. *Solar heating adviser:* N. B. Saunders. *Architect and contractor:* Mark Crosley (son of owner). *Owner and occupant:* Richard Crosley. *Cost of house including solar heating system:* About $52,000. *Cost of solar heating system itself:* $2500. *Funding:* Private.

House employing an 800-sq.-ft. trickling-water-type collector, 1600 gals. of water, and 25 tons of stones

Street   39°N
(25 mi N of Balti-more)
Bay Road

Building   This two-bedroom wood-frame house has a basement and no garage. The first story, with a heated area of 1420 square feet, includes a living room, dining room, kitchen, two bedrooms, and a foyer. The 1070-square-foot second story, which is not heated, includes space for future bedrooms. The basement, also not heated, has a ground-level walk-out on the south side. The walls of the house are insulated with 3½ to 6 inches of fiberglass and the ceiling of the first story is insulated with 6 inches of fiberglass. The window area is modest except for the large (9-by-6-foot) picture window in the south wall of the first story. In winter, storm windows are installed. The house faces 3 degrees west of south.

Collection   The 800-square-foot trickling-water-type collector, made approximately according to the Solaris design of H. E. Thomason, is on the south roof, which slopes 45 degrees. The collector extends on both sides of a large dormer. The heart of the collector is a sheet of corrugated aluminum that has a nonselective black coating. The glazing is single, consisting of single-weight glass panes. The main distribution pipe for water supplied to the collector runs along

Building: 1½-story, 2500 sq. ft. **Collector:** 800 sq. ft., trickling-water type **Storage:** 1600 gals. water and 25-ton bin-of-stones **% Solar-heated:** 70-80 (predicted)

the *lower* edge of the collector; it is a 1-inch-diameter copper pipe. Extending upward beneath each 24-inch-wide panel there is a vertical ½-inch-diameter copper riser pipe carrying water to the upper end of the panel, where a near-horizontal feeder pipe 3/8 inch in diameter distributes water (via 1/16-inch-diameter holes) to the valleys of the black corrugated sheet. The coolant is water, with no antifreeze. The system drains when the centrifugal water pump stops. No heat exchanger is used.

Storage   The 1600 gallons of water is in a horizontal cylindrical steel tank, 16 feet long and 4 feet in diameter, surrounded by 25 tons of 1½-inch-diameter stones in a 20-by-8-foot, 8-foot-high bin in the basement. The bin is made of concrete blocks and is insulated with 1 inch of Styrofoam on the sides and on the top with 6 inches of fiberglass situated between joists of the first-story floor. Hot water from the collector flows first into a small tank situated beside the main tank. The small tank has a capacity of 800 gallons and is 8 by 3 feet by 8 feet high. From here it flows into the upper part of the main tank. Water to be sent to the collector is drawn from the bottom of the main tank. When the rooms need heat, room air is circulated upward through the bin-of-stones by a ¼-hp blower.

Auxiliary Heat Source   Two small wood-burning stoves and an oil-fired furnace.

Domestic Hot Water   This is preheated in a 40-gallon steel tank immersed in the above-mentioned 800-gallon tank. Final heating is by an oil-fired boiler. Note: The collector continues to operate in summer, serving only the 800-gallon tank. The 1600-gallon tank and the bin-of-stones are allowed to cool off. A thermostat prevents circulation of water to the collector when the temperature of the 800-gallon tank exceeds 140°F.

Cooling in Summer   Two small window-type air conditioners are used to cool the two bedrooms.

Designers: P. F. Akehurst and James Burman. *Builder, owner, occupant:*   James Burman. *Cost of solar heating system:*   About $7000. *Funding:*   Private.

Acorn Structures, Inc., House

House employing 480-sq.-ft. water-type collector that is single-glazed with a wavy sheet of plastic

Acton   42½°N
(20 mi NW of Boston)

Building   The collector is on the garage, which is 20 feet west of the house. The single-story, three-bedroom, 1400-square-foot house is of traditional appearance, with an unheated attic and no basement. It is of a standard type produced by Acorn Structures, Incorporated. There is 3½ inches of fiberglass in the walls and 6½ inches in the roof. The garage is 20 by 20 feet and is not heated.

> **Building:** 1-story, 1400 sq. ft.
> **Collector:** 480 sq. ft. water type
> **Storage:** 2400 gals. water
> **% Solar-heated:** 45

Collection   The water-type collector, with gross area of 480 square feet and net area of 420 square feet, is mounted on the 24-by-20-foot south roof of the garage. Its slope is 47½ degrees. There are six panels, of special design. The absorbing structure, designed by Raytheon Company, employs a formed .015-inch aluminum sheet. Copper tubes, 3/8 inch in diameter, 19 feet long, and 5½ inches apart on centers, are recessed into and clipped to this sheet. The headers are 1 inch in diameter. The black coating is nonselective. The coolant is water to which Barclay Algaecide BM has been added. There is no antifreeze: the liquid is drained before freeze-up can occur; draining occurs automatically whenever the 300-watt centrifugal circulation

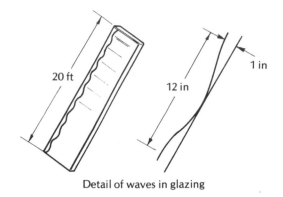

Detail of waves in glazing

pump stops. The circulation rate is 12 gpm. Each panel is glazed with a single 20-by-4-foot sheet of .037-inch Filon, which consists of fiberglass and polyester with a Tedlar coating on the upper surface. The sheet is held in wavy form by wavy edge strips; this form increases transverse stiffness and allows for longitudinal thermal expansion. The panel frame is of wood. The backing includes 2 inches of fiberglass. A silicone sealant is used. The hot water from the collector is delivered directly to the storage tank; there is no heat exchanger.

Storage   The 2400-gallon storage tank, of wood and concrete blocks, is situated beneath the floor at the south end of the house. The tank is 10 by 9 feet by 3½ feet high and has a neoprene liner. The sides and bottom of the tank are insulated with 1 inch of urethane foam and the top is insulated with 2 inches of Styrofoam. Heat is delivered from the storage system to the rooms by a fan-coil system employing a low-power (380-watt) fan. The coil is of especially large size, and accordingly heat extraction from the storage system on a cold night can continue (if necessary) until the temperature of the storage system is reduced almost to room temperature. As a corollary, collection efficiency on the next sunny day is especially high.

Auxiliary Heat Source   Electric heaters in the main hot-air supply duct.

Domestic Hot Water   This is not solar-heated.

Cooling in Summer   No cooling is provided.

Solar engineer, architect, builder, owner: Acorn Structures, Incorporated, John Bemis, President. *Consultant:* Raytheon Company.

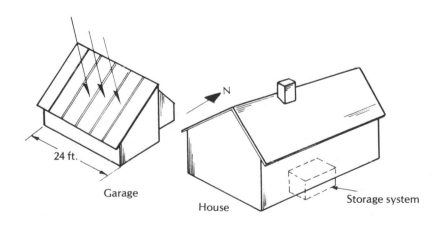

24 ft.

Garage

N

House

Storage system

Experimental building 75-80% solar-heated by passive system employing south windows of unique type, venetian blinds with horizontal aluminized vanes, and a ceiling consisting of 400 sq. ft. of Glauber's-salt-filled tiles

Cambridge   42°N
270 Vassar Street

Building   This one-room, 40-by-22-foot experimental building, with 10-foot inside height, is used as a classroom and drafting room. Because the lot was small and was oriented 20 degrees east of south, the designer was forced to orient the building similarly. Yet by providing four 20-degree-slanted bays he was able to have the four main window areas face exactly south. Each bay is 8 feet long and is displaced 3 feet with respect to the next bay. The windows on the east, north, and west sides of the building are of negligible area and are double-glazed. At the north corner of the building there is an air-lock entryway. The floor is a 4-inch concrete slab resting on gravel. The slab perimeter is insulated by a 4-inch layer of Styrofoam SM which extends 18 inches downward and then runs 36 inches outward. The walls include 6-inch fiberglass batts confined between gypsum boards. Because the 6-inch studs are of steel (to conform to the building code applicable to classrooms) and could constitute a major path for conductive loss of heat, a 1-inch sheet of Styrofoam board, protected by 1/8 inch of stucco cement, has been applied to the exterior of the wall. The horizontal ceiling is supported by 12-inch-deep I-beam joists of corrosion-protected (primed) cold rolled steel, 24 inches apart on centers. The lower flanges of the beams support the thermal storage tiles discussed below. Above the tiles, and between joists, is an 11-inch layer of fiberglass batts. Above this is a steel sheet supporting a 2-inch Styrofoam board covered with conventional tar-and-felt roofing. The R-values of walls and roof are 18 and 33, respectively. The total heat loss of the building has been found by measurement to be 5800 Btu per degree-day. In a twenty-four-hour period when the outdoor temperature is 25°F, the building requires about 40 x 5800 Btu, or about 230,000 Btu. If internal heat sources supply about 15 percent of this, the remaining requirement—to be satisfied by the solar heating system or the auxiliary system—is about 200,000 Btu.

Passive Solar Heating System   This includes four components, each of unique design: south window system, venetian blinds, ceiling, and window seats.

South window system.   Each south-facing bay of the south wall includes a pair of windows, each 37 inches wide and 88 inches high, the sills being 23 inches above floor level. The total area of these windows is 180 square feet, i.e., about 45 percent of the total projected area of the south wall—a small enough percentage so that designers of other such buildings would have some freedom of design of window areas and interior layouts. Each window is triple-glazed with two sheets of glass and one sheet of coated plastic. The glass sheets are of 1/8-inch non-tempered PPG float glass which has a transmittance of 91 percent with respect to solar radiation; these sheets are 1½ inches apart. The coated plastic sheet, midway between these, is a .005-inch mylar sheet, both faces of which have been coated with extremely thin layers of a number of materials so as to provide high reflectance

Building:	1-story, 880 sq. ft.
Collector:	See text
Storage:	
% Solar-heated:	75-80

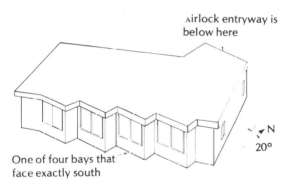

Airlock entryway is below here

One of four bays that face exactly south

N 20°

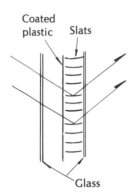

Coated plastic   Slats

Glass

Cross section (looking west) of a portion of a window assembly

with respect to the 4-to-40-micron infrared radiation emitted by objects at about room temperature; with respect to 10-micron radiation, the reflectance is 80 percent. The set of coatings, while providing the desired high infrared reflectance, keeps the reflectance of solar radiation (.3 to 3 microns) low. The transmittance of solar radiation is 70 percent—distinctly lower than that of ordinary glass or plastic sheets but still satisfactorily high. The unique function of this plastic sheet, which the supplier calls Heat Mirror, is to intercept and reflect back into the room the infrared radiation which is continuously emitted by the room contents and tends to escape via the windows. The function is not important on sunny days, because the sky and landscape also emit such radiation and the amount entering the windows (if the special plastic sheet were absent) would approximately equal the amount leaving via the windows; but on cold nights, when the amount of radiation received from the sky and landscape is only 60 to 80 percent of the amount emitted by the room contents, the special sheet is very helpful. Especially because the sheet is situated between the warm inner glass sheet and the cold outer glass sheet, the amount of heat it conserves on cold nights is large. It is said that, typically, the heat loss through such a window is only about one third or one quarter of the loss through an ordinary double-glazed window.

Venetian blind.   Incorporated within each window, in the ¾-inch space between the coated plastic sheet and the inner glass sheet, is a special venetian blind, the slats of which are aluminized and are tilted (nearly horizontally, typically) so as to reflect direct midday solar radiation upward toward the ceiling of the south half of the room. The general design is much like that of a standard venetian blind made by the same supplier (Rolscreen Company of Pella, Iowa). Each slat is 38 inches long, ¾-inch wide, and .01 inch thick. The cross section is very slightly curved, with the concave face upward—the reverse of the orientation used in most venetian blinds. The curvature increases the stiffness and also improves the optical performance of the blind as a whole. The concave face of the slat is coated with a film of aluminum on mylar and has, initially, a reflectance of about 85 percent. The slats are 0.4 inch apart—an unusually close spacing. They are supported by three vertical "ladders" spaced 14 inches apart. The vertical members of the ladders are 1/16-inch-diameter polyester cords and the cross members (rungs) are of 1/64-inch polyester strings. The rungs are in pairs, with the slats positively confined by them. Every few weeks the tilt of the slats is changed slightly, by adjusting a knob, to conform to the change in altitude of the sun at midday. (There is no need to adjust them during the course of any one day, inasmuch as altitude of the sun changes by less than 6 degrees during the central five-hour period of the day and the windows face exactly south.)

The window system as a whole, with its sequence of glass, plastic, reflective venetian blind, and glass, has a transmittance of about 50 percent—computed by multiplying together these four values: .91, .70, .85, .91.

Ceiling.   The south half of the 10-foot ceiling is faced with energy-storing tiles which receive solar radiation that has been reflected upward by the venetian blinds. Each tile is 23½ inches square and 1¼ inches thick. The one hundred tiles are arranged in rows defined by the above-mentioned ceiling joists, which are 24 inches apart on centers; the tiles are supported, at the edges, by the 1-inch-wide flanges constituting the lower faces of the joists. No fastenings are used and any tile can be removed quickly, without use of tools, if this is desired—to permit inspection, for example. The total area of the ceiling tiles is approximately 400 square feet. (The north part of the ceiling also is covered with tiles, but these receive little radiation and are for a special experimental purpose not discussed in this account.)

Each tile consists of a case and, within it, two side-by-side pouches. The case is of polymer concrete that consists of 85 parts by weight of small-stone aggregate, 15 parts of polyester resin, and a few parts of fiberglass. The top and bottom are ¼ inch thick and the sides and ends are ½ inch thick. The material is waterproof but is very slightly permeable to water vapor—which is the reason that impermeable pouches are used also. Each pouch, which contains 12 pounds of phase-change material (PCM), is 22½ by 11 inches and ¾ inch thick. It is made of .002-inch aluminum that is sandwiched be-

tween two .002-inch polyethylene films. A horizontal septum of .004-inch polyethylene divides the space within the pouch into two horizontal regions, each of which is thin enough (3/8 inch) so that no irreversible settling-out of the more dense components of the PCM will occur. All edges of the pouch and of the septum are heat-sealed.

The PCM is a special formulation designed to melt and freeze at 73°F (plus or minus about ½°), to have a large latent heat of fusion, to freeze (when heat is extracted) without supercooling, and to melt (when heat is added) without superheating. The formulation includes 85 percent by weight sodium sulfate dekahydrate (Glauber's salt, $Na_2SO_4 \bullet 1OH_2O$), together with about 9 percent sodium chloride (NaCl), which helps reduce the phase-change temperature toward the desired value of 73°F, 3 percent borax ($Na_2B_4O_7$), which helps initiate crystal formation when heat is extracted and thus helps avoid supercooling, and 3½ percent of Cab-O-Sil, a thickening agent that contains fumed silica. This last material, having a rigid but open three-dimensional structure, is capable of engaging and immobilizing large quantities of water or aqueous solutions. Thus when the PCM receives heat and melts, the heavier components of the liquid do not descend and the lightest component (the water of crystallization released by the Glauber's salt) does not ascend. The material as a whole remains homogeneous, ready to recrystallize in the normal way when heat is extracted. Whereas pure Glauber's salt melts and freezes at about 88°F, the formulation used here melts and freezes at much lower temperature: 73°F, selected as being ideal for the purpose at hand. Although pure Glauber's salt is said to have a latent heat of fusion in the neighborhood of 100 Btu per pound, the value may be reduced by various additives and by various methods of temperature cycling; some of the changes are not fully understood. A relatively low value—33 Btu per pound—has been taken as appropriate to the present formulation.

Each tile has a total weight of 44 pounds, of which about 24 pounds is attributable to the PCM. Thus the latent heat of the PCM in one tile is about 790 Btu. The case contributes slightly to heat storage; if its typical temperature change is 10 Fahrenheit degrees, it may increase the storage capacity by about 50 Btu per tile, making a total of about 840 Btu per tile.

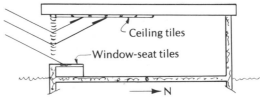

Cross section (looking west) showing window, venetian blind, and energy storage tiles

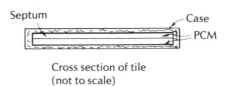

Cross section of tile
(not to scale)

The underside of the tile is bluish-black and absorbs about 85 percent of the solar radiation incident on it from below. Heat travels upward within the polymer concrete and into the PCM. When the radiation is intense, as at noon on a sunny day, the rate of delivery of energy to the tile may be so great that there may be a 10-degree temperature difference across the liquid PCM film beneath the solid PCM layer; the difference, although undesirable, is not great enough to be significantly harmful.

At all times the tiles emit 4-to-40-micron infrared radiation in all downward directions and at the same time they receive such radiation traveling upward from the floor, the walls, furnishings, etc. On most nights in winter the tiles are hotter than these objects and accordingly the downward flow of radiant energy exceeds the upward flow; thus the tiles keep the room temperature fairly high—close to 70°F. Operating experience gained in midwinter of 1978 indicates that, ordinarily, room temperature measured at locations 5 feet above the floor seldom falls below 62°F at night and seldom rises above 72°F during the day. Because the rate at which the tiles give off heat at night is far lower than the rate at which they receive heat at midday, the temperature differences that occur within the tile during the night are only a small fraction of the 10-degree difference mentioned above.

Window seats. Adjacent to each of the four large window areas there is a 7½-by-3-foot horizontal array of energy storage tiles. Each such array is about 23 inches above floor level, i.e., at about the same level as the base of the window, and may be used as a window seat. There are eight tiles in each such array, and 32 tiles in all; their aggregate area is 90 square feet. Each such tile has slightly smaller area than the ceiling tiles have, but has the same thickness. The internal designs are the same and the materials used are the same. If the lower parts of the venetian blinds are raised slightly, solar radiation strikes the window-seat tiles directly and warms them. Thus they can contribute to keeping the room warm at night by radiation and convection. Also, they provide a warm place to sit. Their dark color helps reduce glare here.

The total effective thermal capacity of the ceiling tiles and window-seat tiles is about 125 times the figure presented above for a single ceiling tile, 850 Btu, or about 106,000 Btu. Of course, the concrete floor and the walls also contribute to the thermal capacity of the building as a whole.

Auxiliary Heat Source This consists of 6 kw of electric baseboard heaters.

Domestic Hot Water None.

Cooling in Summer There is no formal cooling system. The cooling requirement is small, inasmuch as the roof and walls are well insulated, the venetian blinds may be adjusted to exclude solar radiation, and the infrared-reflecting plastic sheet prevents influx of such radiation from the surrounding air and landscape. Windows and vents, including four 7½-by-2-foot vents in the ends of the south bays, may be opened to provide cross drafts and to allow cool air to enter at night. The thermal masses of the ceiling and the concrete floor help keep the daytime temperature rise small.

Comment on General Performance   The system operates with no moving parts, no noise, no adjustments (except adjustment of the venetian blinds every few weeks), and no dependence on electricity. It avoids several drawbacks of earlier types of passive solar heating systems employing large window-walls: near the window-walls of such systems the occupants may find the glare to be excessive by day and may feel cold at night; also, much heat is lost through those walls at night. In the present system, the venetian blinds and dark-colored window seats greatly reduce glare (but without sacrifice of view, because the slats are nearly horizontal); the warm window-seats eliminate chill; and the third glazing sheet, with its infrared-reflecting coatings, greatly reduces heat loss. The durability of the present system has not been fully demonstrated, but thermal cycling tests of the PCM-filled pouches have been encouraging: by the end of February 1978 several such pouches had been heated and cooled two thousand times, melting and refreezing the 3/8-inch layers of PCM each time, with no detectable change in performance.

Problems and Modifications   In March 1978 some of the south windows lacked the intermediate infrared-reflecting plastic glazing sheet because the supplier had encountered production difficulties.

General concept: T. E. Johnson of the Department of Architecture, Massachusetts Institute of Technology; also Sean Wellesley-Miller and Day Chahroudi. *Solar design and general design:*   T. E. Johnson, Crisp Benton, and Stephen Hale of the Massachusetts Institute of Technology. *Supplier of the infrared-reflecting plastic sheet:*   Suntek Research Associates (esp. Day Chahroudi and Sean Wellesley-Miller). *Supplier of the special venetian blinds:*   Rolscreen Company. *Supplier of the tile cases:*   Architectural Research Corporation. *Supplier of the PCM mixture:*   Cabot Corporation (which is the producer of Cab-O-Sil). *Source of funds for materials and construction:*   Godfrey L. Cabot Research Fund at MIT. *Funding of instrumentation, data recording, and data analysis:* $72,000 from the U.S. Department of Energy.

3200-sq.-ft. house 85% solar-heated by 800-sq.-ft. collector on remote garage

Concord   42½°N
(suburb of Boston)
330 Garfield Road

Building   This four-bedroom house has a half-basement and crawl space, and no attic. The walls are insulated with 3½ inches of fiberglass and the ceilings contain 6 inches of fiberglass. The area of the south windows is 400 square feet and the area of the other windows is 200 square feet. All of the windows are double-glazed. Part of the first-story floor is of wood and part is of tiles on concrete. Some use is made of insulation on the outside of the basement walls. Some north rooms can be isolated and left unheated in cold midwinter periods.

Building: 2½-story, 3200 sq. ft.
Collector: 800 sq. ft., water type
Storage: 4000 gals. water
% Solar-heated: 85

Collection   The collector is on the south roof of the two-car garage, which is 60 feet northwest of the house. The roof slopes 60 degrees. The collector is 36 feet long and 24 feet wide. It consists of fifty-four experimental panels made by Kennecott Copper Corporation, each 8 feet by 22 inches and arranged in three rows. The absorber is a .004-inch sheet of copper to which rectangular-cross-section copper tubes are soldered 2 inches apart on centers. The inside dimensions of the tube cross section are: width, ½ inch, height, 1/8 inch. The black coating is nonselective. The panels are double-glazed; the outer sheet is .032-inch Filon and the inner sheet is .004-inch Tedlar. The backing includes 4 inches of fiberglass. The coolant is demineralized,

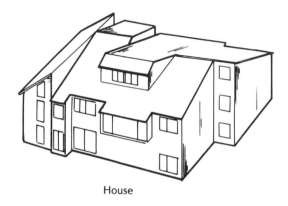

Garage (60 ft NW of house)

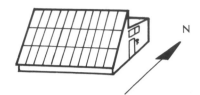

N

House

pH-controlled water. No antifreeze is used; the liquid is drained automatically whenever the 1-hp centrifugal circulation pump stops. There is no heat exchanger; the hot water from the collector flows directly into the storage tank. Insulated copper pipes, 1½ inches in diameter, run underground between the garage and the house. Flow starts whenever the collector temperature exceeds the storage tank temperature by 10 Fahrenheit degrees and stops when the two temperatures become equal.

Storage The 4000-gallon steel-reinforced concrete storage tank is situated in the southeast corner of the house. The sides of the tank— but not the bottom—are insulated externally with 4 inches of Styrofoam. The house has two heating zones, served by two fan-coil systems. The coils receive hot water from the storage tank; a ¼-hp centrifugal pump circulates the water.

Auxiliary Heat Source An oil-fired boiler, a wood-burning stove, and two fireplaces.

Domestic Hot Water This is not solar-heated.

Cooling in Summer The above-mentioned tank can be filled with cool (50°F) water from a well and on the hottest days this water can be circulated to the fan-coil systems.

Solar engineer, owner, occupant: Richard Thornton. *Architect:* Day and Ertman. *Builder:* Acorn Structures, Incorporated, and Donald Locke.

House built in 1955 and retrofitted for solar heating in 1976

Hingham 42°N
(suburb of Boston)
Cushing Pond area

Building This is a three-bedroom wood-frame house. The full base-ment includes several rooms and, at the northwest corner, a one-car garage. There is no attic. The insulation conforms only to 1955 stan-dards. The house faces straight south, and on the south side the win-dow area is larger than on the other three sides. All windows are dou-ble-glazed. A 12-by-9-foot greenhouse is attached to the south side of the main story.

Collection The collector is mounted on the horizontal roof and consists of Daystar panels, with a slope of 45 degrees, in two east-west rows 16 feet apart on centers. For heating, there are sixteen panels, each 73 by 44½ inches and 5½ inches thick. The heart of the panel is a .020-inch copper sheet coated with nonselective black 3M Nextel. Brazed to the back of the sheet is a serpentine ½-inch-ID copper tube with the segments 6 inches apart on centers. Between the copper sheet and the glazing—a single sheet of 3/16-inch tem-pered glass—is a 2-inch airspace that is filled with accordion-pleated .005-inch polycarbonate film, which helps to trap reradiation-band radiation and which, more importantly, inhibits air convection. (The reflection loss of this pleated film is well below the approximately 8 percent loss of a planar film.) The panel is backed with 2 inches of isocyanurate foam. All panels are hydraulically in parallel and are supported by pipes bolted to the roof. The water-glycol coolant is cir-culated, at a maximum rate of 16 gpm, by two TACO centrifugal pumps connected in series so that if one pump fails the other will maintain some circulation. Two Daystar heat-pump panels operate automatically to keep the collector from becoming too hot.

Storage Buried one foot underground just to the west of the house is the main tank, a horizontally mounted cylinder 4 feet in diameter and 6 feet long, made of 7-gauge steel and insulated with 3 inches of polyurethane foam and tar-paper wrapping. There is also a 15-gallon expansion tank. Within the main tank is a heat exchanger employing

Building: 2-story, 1650 sq. ft.
Collector: 330 sq. ft., water type
Storage: 500 gals. water
% Solar-heated: 70

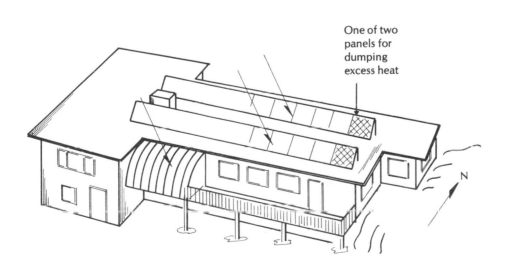

One of two
panels for
dumping
excess heat

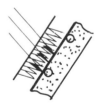

Cross section of
portion of
collector

113

five finned coils in parallel. Circulation to the collector starts when ΔT exceeds 10 Fahrenheit degrees relative to storage. Rooms are heated by radiators served by hot water either from the solar storage system (never directly from the collector) or from the auxiliary heater, as determined by a logic panel. The house has three heating zones.

Auxiliary Heat Source   Oil-fired boiler, 192,000 Btu/hr. (previously installed).

Domestic Hot Water   Heated by solar heating system, supplemented if necessary by the above-mentioned heating boiler. There are also electric coils as extra backup.

Cooling in Summer   Conventional 4½-ton air conditioner (previously installed), made by Space Conditioning, Inc.

Problems and Modifications   At night, a reverse flow of water, from the storage system to the collector, sometimes occurred, with considerable loss of heat. This was corrected by installing a check valve. Snow sometimes sticks to the collector glazing and must be removed by shoveling or brushing, or by melting it by circulating a little hot water from the storage system through the collector.

Solar engineering and installation: New England Solar Systems, L. Mazzini, President. *Owner, Occupant:*   Richard L. Katzenstein. *Cost of retrofit:* $16,500.

Retrofit system employing 960-sq.-ft. water-type collector and 2000 gals. water in a plywood tank

Lincoln   42°N
(suburb of Boston)
At Drumlin Farm

Building   Solar heating was applied to an old (1951) two-portion structure consisting of (1) a nature center, a 70-by-40-foot, three-story wood-frame building on the old foundations of a barn, and (2) a gift shop, a 48-by-30-foot, 1½-story wood-frame building adjacent to the southwest corner of the nature center. The main floor of the shop is on the same level as the lowest floor (basement) of the nature center. Both buildings have moderate window areas and the windows are double-glazed. The walls of the nature center are insulated with 3½ inches of Borden urea formaldehyde and the walls of the shop are insulated with 3½ inches of fiberglass. The roofs are insulated to R-19. The uppermost 2-foot portion of the shop foundation wall is insulated on the outside with 2 inches of Styrofoam board. The roof of the shop is new and was designed especially to accommodate a collector.

Collection   The water-type collector, which has a gross area of 960 square feet and a net area of 840 square feet, is mounted on the new 45-degree roof of the gift shop, which faces 15 degrees east of south. The collector is integral with the roof and rests on roofing paper and plywood sheathing. The twelve collector panels, each 20 by 4 feet, are of Acorn Structures, Incorporated, design. The aluminum sheet has a nonselective black coating, and eight 3/8-inch-diameter copper tubes are tightly embraced by formed channels, or grooves, in this sheet. The glazing is single and consists of Filon: a Tedlar-coated sheet of polyester and fiberglass. The sheet is held in slightly wavy form in order to increase the transverse stiffness and allow for longitudinal (vertical) thermal expansion. The backing includes 2 inches of fiberglass. The coolant, which is water with no antifreeze, is circulated, via 1½-inch-diameter feed pipes, at 24 gpm by a ⅓-hp centrifugal pump. The system drains automatically whenever the pump stops, which occurs whenever the level of radiation drops so low that the collector cannot operate productively. The collector manifolds, built into the panels, are pitched to drain. The hot water from the

Building: See text
Collector: 960 sq. ft., water type
Storage: 2000 gals. of water
% Solar-heated: 50-60 (predicted)

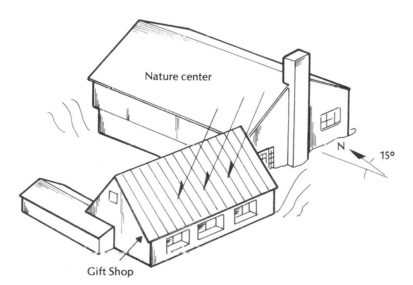

Nature center

N   15°

Gift Shop

collector flows directly into the storage tank; there is no heat exchanger.

Storage   The 2000 gallons of water is contained in a vertical cylindrical tank 7½ feet in diameter and 7 feet high. The tank is assembled from curved sheets of 3/8-inch plywood held together by galvanized steel straps. A .030-inch vinyl liner is used, and 3½ inches of fiberglass is applied to the outside of the tank. The heat distribution system has two zones, serving shop and center. When the rooms of the shop need heat, water from the storage tank is circulated through a total of 230 feet of finned-tube baseboard radiators in these rooms. The tube lies a few inches below another finned tube that is served by the auxiliary heating system. The hot-water supplies for the two tubes are independent and the control systems likewise are independent. When the storage tank is very hot, the tube it serves keeps the rooms warm. When it is not very hot, the finned tube it serves continues to operate, supplying a fraction of the heat needed (and cooling the tank to, say, 80°F), and the finned tube served by the auxiliary system operates also, in parallel. Because the two kinds of tubes are situated close together in the same gravity-convective airstream, each augments the speed of airflow past the other. The thermostats used are set so as to make the system as a whole keep the room temperature at, say, 70°F during the day and at 55°F at night. The nature center also is partially heated by the solar heating system by means of a finned-tube loop.

Auxiliary Heat Source   Existing oil-fired boiler.

Domestic Hot Water   This is preheated in a 190-foot coil of ½-inch-diameter copper tubing within the storage tank. Percent solar-heated: 75 (predicted).

Cooling in Summer   None. Little cooling is needed, inasmuch as the building is well insulated and the south windows are recessed so deeply that, in summer, little direct radiation can enter.

Problems and Modifications   After the solar heating system had been in use for a few months, a "cosmetic" defect in the collector was apparent. The initially arranged wave pattern in the plastic cover sheet of the collector became distorted, presumably because the sheets had been shipped and stored in roll form and had some "memory" of that condition. The sheets were removed and were replaced with sheets that had never been rolled up and thus readily accepted and held the desired wave pattern.

During the first few months of operation the storage tank lost many gallons of water by evaporation. When water was added to restore the usual water level in the tank, the tank was cold and a little too much water was added; when, later, the tank warmed up, the water expanded enough so that a few gallons overflowed onto the basement floor. To prevent recurrence of loss of water by evaporation, the maintenance staff installed a condensation trap.

Solar engineers and architects: Massdesign Architects and Planners (esp. T. G. Ingersoll and G. F. Tully). *Collector designer:*   Raytheon Company (esp. Will Hapgood).

Manufacturer of collector: Acorn Structures, Incorporated (esp. John Bemis) *Solar consultant:* Solar Heat Corporation (esp. Mark Hyman). *Structural engineer:* Souza and True. *General contractor:* H. Tobiason Builders, Incorporated. *Heating contractor:* Ken-Mar Corporation. *Owner, occupant, and source of funds:* Massachusetts Audubon Society. *Cost of solar heating system proper:* $30,000 to $40,000. *Cost of instrumentation for monitoring and display:* $13,000.

Massachusetts Audubon Society Gift Shop. At 9:15 a.m. on Jan. 8, 1977, most of the collector was covered by heavy snow that had fallen during the night.

Fifteen minutes later, most of the snow on the collector slid down, clearing the collector for action. The observer (photographer T. G. Ingersoll) was amazed—inasmuch as the ambient temperature was well below freezing, the buildings were locked, no employee had yet arrived, and no one had touched the solar-heating-system controls. Presumed explanation: the sensor that controls circulation of water from the storage system to the collector is at the extreme top of one panel; thus as soon as the top became free of snow, the sensor became hot and triggered a flow of warm water to the collector. This quickly melted the under-surface of the snow, and all the snow slid off.

117

Brickelmaier House

North Chatham 42°N
(on Cape Cod)
32 Cranberry Lane

Building: 2½-story, 1400 sq. ft.
Collector: 480 sq. ft., water type
Storage: 2000 gals. water
% Solar-heated: 70

House employing 480-sq.-ft. water-type collector and 2000 gallons of water in vinyl-lined wooden tank

Building   This is a two-bedroom, wood-frame, Cape-type house with loft, small attic space, and a full basement that has walk-out patio doors. The window area is 314 square feet, of which 30 percent is on the south side. All windows are double-glazed. The walls are insulated with 3½ inches of fiberglass and the ceiling or roof contains 6½ inches of fiberglass.

Collection   The collector, on the south roof sloping 47 degrees, is of Acorn Structures, Incorporated, type and consists of a single row of six panels, each 20 feet high and 4 feet wide. The absorbing structure, of Raytheon design, employs a formed .015-inch-thick aluminum sheet with a nonselective black coating. Copper tubes 3/8 inch in

diameter and 19 feet long are clipped to this sheet, 5½ inches apart. The headers are 1 inch in diameter. The coolant is water to which Barclay Algaecide BM has been added. No antifreeze is included; the system drains automatically whenever the 300-watt centrifugal pump stops. The glazing is a single 20-by-4-foot sheet of .025-inch Filon (polyester and fiberglass). The sheet is held in wavy form by wavy edge support strips to increase stiffness and allow for thermal expansion. The panel frame is of wood. The backing includes 2 inches of fiberglass. A silicone sealant is used.

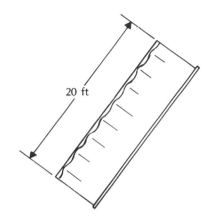

Storage   The 2000 gallons of water is in a vinyl-lined wooden tank in the basement. The tank is 8 feet in diameter and 6½ feet high. It is insulated with 3 inches of fiberglass. Heat is delivered to the rooms by a fan-coil system that employs a low-power (380-watt) fan.

Auxiliary Heat Source   Oil furnace and forced-air system.

Domestic Hot Water   This is preheated in a 30-gallon tank within the main tank.

Cooling in Summer   None.

Problems and Modifications   Initially, a pipe fitting attached to the storage tank was installed incorrectly and some water leaked out. This was soon corrected. On some occasions when the storage tank was especially hot a significant amount of water evaporated from the surface and some moisture found its way into the house proper. The difficulty was overcome by installation of a vent cap.

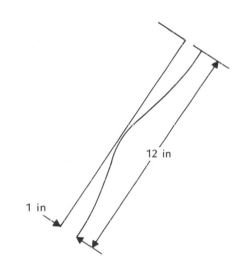

Detail of waves in glazing

Solar engineering, general design, architect, fabricator: Acorn Structures, Incorporated (John Bemis, et al.). *Builder:* Cape Associates Incorporated (L. F. Langhans, Jr., et al.). *Owner, occupant:* June Brickelmaier. *Funding:* Private.

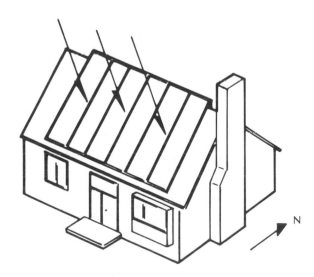

Princeton   42°N
(15 mi NW of
Worcester)
Worcester Road
(Route 31)

Building: 1½-story, 1500 sq. ft.
Collector: 800 sq. ft., trickling-water type
Storage: 1600 gals. water, 30-ton
          bin-of-stones
% Solar-heated: 65

House employing water-and-bin-of-stones storage system

Building   The wood-frame house is 40 by 24 feet and has three bedrooms, a small attic, and a full basement. Walls are insulated with 3½ inches of fiberglass, and the ceiling or roof with 6 inches. The house faces 10 degrees west of south, and on the south facade there are four pairs of sliding glass doors. These are double-glazed, as is the small window area.

Collection   The collector occupies most of the 24-by-40-foot south roof, which has a slope of 50 degrees. Each of the nine panels, made by The Mountain Company, is 24 by 4 feet and contains a .021-inch sheet of corrugated aluminum coated with nonselective black Sherwin-Williams Polane. Glazing is a single .040-inch sheet of Kalwall Premium Sun-Lite. When collector temperature exceeds storage tank temperature by 10 Fahrenheit degrees or more, a ⅓-hp centrifugal pump circulates water by way of a 1-inch-diameter copper pipe along the ridge of the roof to the panels. At the top of each panel is a 3/8-inch-diameter manifold, with one 1/16-inch hole for each valley of

the corrugated sheet, by which the water is fed into the collector. No inhibitor or antifreeze is used; water is drained from the roof before freeze-up can occur.

Storage   Water is stored in a horizontal, cylindrical steel tank 4 feet in diameter and 17 feet long, which is not insulated but is surrounded by 2-inch diameter stones in a concrete bin, 22 by 9 feet by 7 feet high inside, located in the west end of the basement. The exterior of the bin is insulated with 3½ inches of fiberglass. When rooms need heat, room air is circulated through the bin-of-stones and ducts by a ¼-hp blower.

Auxiliary Heat Source   Oil-fired hot water heater. The hot water is circulated through a coil situated at the entrance to the duct that carries air to the rooms.

Domestic Hot Water   Water is preheated in a small tank inside the 1600-gallon storage tank described above.

Cooling in Summer   None.

Problems and Modifications   Some water that condensed on the underside of the glazing escaped from the lower end of the collector panels, and an appreciable amount of water was lost from the collector. The problem was largely solved by modifying the closures at the lower ends of the panels. Some of the black paint on the corrugated aluminum sheets flaked off. The auxiliary heating system did not work well at first, but the problem was solved by transferring the heating coil to the entrance of the pertinent duct. The architect believes it would have been desirable to have used much thicker insulation on walls and roof.

Solar engineer, architect, owner: W. E. S. Bird. *Builder:*   The Mountain Company. *Cost of house alone:* $45,000. *Cost of solar heating system:* $5000.

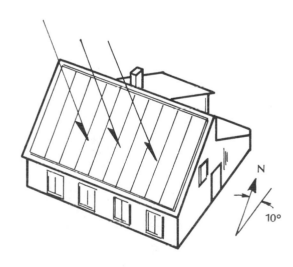

Solar Heat Corp. House 2

Waltham   42½°N
(suburb of Boston)

Building: 1½-story, 2300 sq. ft.
Collector: 1200 sq. ft., water type
Storage: 16,000 gals. water
% Solar-heated: 95

With very large [1200 sq. ft.] water type collector and very large [16,000 gals.] water storage system, this house is 95% solar heated.

Building The wood-frame Cape-type house has three bedrooms and is 62 feet by 24 feet. The full basement includes a garage at the east end. The house is insulated with fiberglass to R-11 for walls and R-19 for ceilings and roof, and all windows are double-glazed. The house faces 11 degrees west of south.

Collection The collector is located on the roof, which slopes 50 degrees, and is made up of 55 specially designed panels that were assembled in the garage of the house before installation. There are three rows of panels: two lower rows of 21 panels (8 by 3 feet) each and an upper row containing 13 panels 6 by 3 feet. The heart of each panel is a sheet of Olin Brass Company Roll-Bond aluminum with a nonselective black coating. Each sheet has eleven parallel integral passages for water and, at top and bottom, sloping, sure-draining header passages. The panel is glazed with a single sheet of ¼-inch Plexiglas which is ½ inch from the black aluminum sheet and is backed with 3½ inches of fiberglass and 1 inch of Styrofoam board. Untreated, filtered water, without corrosion inhibitor or antifreeze, is used as coolant. To control corrosion, a "getter"—a roll of aluminum screening—is employed in each of three of the header pipes. The water is drained from the collector before freeze-up can occur. Coolant is circulated through the three circuits of the collector and to the storage system at 15 to 20 gpm by a ½-hp centrifugal pump. The two or three panels of any one column are hydraulically in series but

are separated by 5-inch airspaces to allow room for flexible hose connections and for thermal expansion.

Storage: The water is contained in a poured concrete tank, 24 by 15 feet and 6 feet high, that is actually a portion (two fifths) of the basement with a suitable partition. The floor of this portion is 2 feet lower than the rest, so that the partition projects only 4 feet above the main floor and requires no additional strengthening. The tank is insulated inside with 6 inches of Styrofoam and has a vinyl liner. When rooms need heat, water is circulated from this tank to baseboard radiators, which are about twice the normal length to permit keeping rooms warm even when storage tank temperature is fairly low. There are two circulation circuits, each served by a 1/12-hp centrifugal pump.

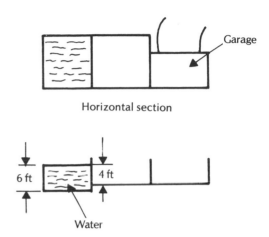

Horizontal section

Vertical section

Auxiliary Heat Source Little auxiliary heat is needed, although during the especially cold winter of 1976-1977 the house was only 92 percent solar-heated. There is a high-efficiency, all-metallic Majestic fireplace in the living room. An 80-gallon tank with a 9-kw electric heater, located in the basement, can be used to raise the temperature of water circulated to the baseboard radiators.

Domestic Hot Water Water is preheated by a coil in the main storage tank and final heating is performed in another 80-gallon tank by a 4.5-kw heater. The solar heating system provides 87 percent of the water heating.

Cooling in Summer None provided.

Problems and Modifications On a few occasions during the winter of 1976-77 when the auxiliary heat source was in use, air blockage occurred in the radiator system. To prevent this, a higher-pressure (but lower-powered) circulating pump was installed late in 1977.

Solar engineering and owner: Solar Heat Corporation, M. Hyman, President. *Architects:* G. Tully and H. A. Russell of Massdesign Architects and Planners. *Builder:* E. Tobiason Builders, Inc. *Occupant:* Mark Hyman.

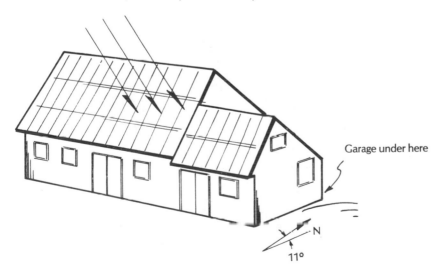

Garage under here

Cambridge School Solar Building

Weston
(suburb of Boston)
Lexington Street

Building:	2-story, 14,400 sq. ft.
Collector:	} See text
Storage:	
% Solar-heated:	25 (predicted)

Passive solar heating system employing stepped solar roof and massive walls deep within the building

Building   The east end of this school building was damaged by a fire and was then remodeled and rebuilt. The west end is new. Late in 1975 solar heating was applied to two thirds of the building. The building faces 26 degrees west of south. It is 160 feet long, and its east, central, and west portions are 36, 48, and 56 feet wide respectively. The upper of the two stories—the main story—contains a kitchen and a large dining room. The lower story, a basement, contains classrooms and utility rooms. The total floor area is 14,400 square feet. Some north walls and transverse walls are 12 inches thick and are of concrete blocks having 25 percent voids. The massive exterior walls are insulated on the outside with 4 inches of fiberglass protected by a layer of stucco. (Note: The masonry walls are required by the fire code but serve also to store thermal energy.) The basement floor is of concrete. A major portion of the vertical south wall consists of windows, which provide view and admit solar radiation. About half of the window area is single-glazed and half is double-glazed.

Passive Solar Heating System   Solar radiation enters the upper part of the building via a special roof area, or large skylight, serving the central and western portions of the building. The dimensions of the

special roof are 96 by 18 feet; its area is 1700 square feet. The slope is 25 degrees. The roof consists mainly of a staircase-like structure. The "risers" are 8½ inches high and consist of a single sheet of double-strength glass. The "treads" are 18 inches wide and are of shiny aluminum. Resting on this stepped structure there is a sloping, corrugated, translucent sheet that sheds rain and snow and—together with the "staircase"—defines a dead-air space that provides some insulation. This sheet is of Filon, which contains polyester and fiberglass. In winter much radiation passes straight through the sloping cover and the risers and penetrates deep into the rooms. Also, much radiation strikes the aluminum treads and is reflected by them, then passes through the risers and strikes the undersides of other treads, and is reflected obliquely downward into the rooms. Storage is provided by the building as a whole and especially by the above-mentioned massive walls, many of which receive direct solar radiation.

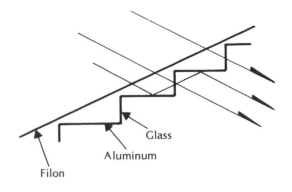

Cross section of special roof

Auxiliary Heat Source　Oil-fired hot water system.

Domestic Hot Water　This is not solar-heated.

Cooling in Summer　In summer the special roof excludes most of the solar radiation incident on it: the rays come from steeply overhead and are reflected by the aluminum treads back up to the sky. The massive walls and floors keep the rise in indoor temperature small and are cooled at night by circulation of outdoor air.

Note Concerning Illumination　Much daylight enters the rooms via the special roof. Even deep interior parts of the main story require little or no artificial illumination on clear days in summer and winter. The amount of money saved is substantial. Unlike most passive solar heating systems, this one does not entail excessive glare in the south parts of the building.

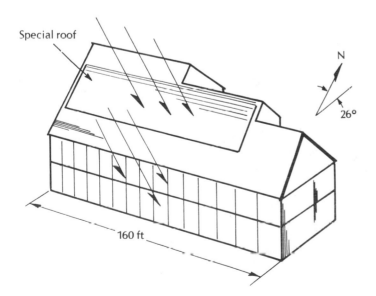

Instrumentation Temperatures at sixteen locations are recorded on-site on magtape. The analysis of the data is made at a remote laboratory specializing in analysis.

Problems and Modifications In cold weather much heat is lost through the special roof. The net gain, via the special roof, in a typical twenty-four-hour period in winter is modest. It is believed that installation of an additional sheet of glazing close beneath the staircase-like structure would greatly increase the net gain.

Solar engineer: N. B. Saunders. *Architect:* Davies, Wolf and Bibbins. *Owner:* Cambridge School. *Cost of the special roof:* Little different from the cost of a conventional roof. *Cost of the temperature sensing and recording system:* $1000. *Funding of computer-assisted analysis of performance data:* $10,000 furnished to N. B. Saunders by the Energy Research and Development Administration in mid-1976; the resulting reports are identified by the designation P.O. Wa-76-4947.

Two-story passively heated house in its 18th year

Weston   42½°N
(suburb of Boston)
15 Ellis Road

Building This is a four-bedroom house with no basement or attic. Earth encloses the north part of the first story. The roof is nearly horizontal, sloping 3° downward toward the north. The walls and first-story floor are very massive, as explained below. The roof, which includes nineteen layers of various materials, has an R-value equal to that of 13 inches of fiberglass. Most windows are double-glazed. The construction of window frames and walls is such as to minimize leakage of air and minimize metallic, conducting, heat-leakage paths.

Building: 2-story, 2600 sq. ft.
Collector: }
Storage: }   Passive
% Solar-heated: 65

Collection Radiation is received through the south window-walls of the first and second stories, the total south window area being 450 square feet. The second-story window-wall, which serves the living and dining areas, consists of six panels, each 6½ by 5 feet and each double-glazed with glass. South of this glass wall there is a 6-foot-deep balcony with 6-foot eaves and 6-foot end-baffles, which control wind and sun. The first story has similar glazing except that an additional glass wall has recently been installed at the south limit of the balcony; this added glazing, which consists of a single layer of 1/16-inch-thick panes, 24 by 16 inches, slopes 80 degrees.

Storage The main storage elements are the massive floors and walls. The first-story floor is a 6-inch concrete slab resting on masonry in which there are nineteen major air channels and a larger number of minor channels. Part of the south portion of the second-story floor also is massive. The walls are of 1-foot-thick pumice blocks; besides providing insulation, they store much heat. The thermal capacity of the system is 75,000 Btu per Fahrenheit degree. Solar radiation strikes

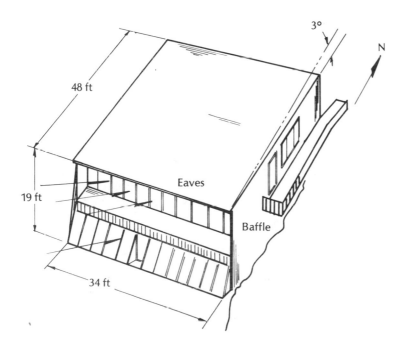

and heats the south portions of the floors and walls, and the open interior design of the house facilitates circulation of warm air to the north rooms. When needed, any of three fan-and-duct systems may be brought into play, manually, to increase general circulation. A fourth fan may be turned on to circulate room air. The fans are of 24-inch diameter.

Auxiliary Heat Source   Electric—chosen because it permits easy evaluation of the amount of auxiliary heat used. The heating elements are concealed within the ceilings.

Domestic Hot Water   This is not solar-heated.

Cooling in Summer   Almost no cooling is needed. Solar radiation is excluded by the eaves, balcony, and baffles. The insulation is excellent. The thermal capacity is large. If rooms tend to become too hot, cool air from beneath the first-story floor is circulated to the rooms.

Problems and Modifications   The building has been an experimental one. The owner-occupant has tried out a variety of collection schemes, including a near-horizontal rooftop collector. Most of the experiments were instructive. Several were disappointing. The present system has performed excellently. A minor fault is that the first-story concrete slab and the masonry below it are in direct contact with wet ground and lose some heat to it.

Engineer, owner, occupant: N. B. Saunders. *Funding:*   Private.

Small experimental building that has a stepped solar roof and a large array of water-filled troughs suspended close below it

Weston   42½°N
(suburb of Boston)
Lexington Street, on
grounds of Cambridge School

Building This is a one-room, 12-by-12-foot experimental building, with a small attic space. There is no basement or foundation: the building rests on beams laid on the ground. The walls and floor are insulated with 3½ inches of fiberglass. There is only one window: a door-size window, at the center of the south side, that is double-glazed with .025-inch Kalwall Sun-Lite. The monoslope roof is 38 degrees from the horizontal. The building faces exactly south.

Building: 1-story, 144 sq. ft.
Collector: }
Storage: } Passive
% Solar-heated: See text

Collection This employs a special roof and a special storage system. The special roof, which slopes 38 degrees, is 12 feet wide and 16 feet high. The gross area is about 180 square feet and the net area is 162 square feet. The roof consists of three 16-by-4-foot panels, each of which includes (1) a translucent sloping cover of .025-inch Kalwall Sun-Lite, which sheds rain, snow, etc., and (2) a staircase-like structure consisting of an alternating series of "risers" and "treads," which run east-west. The risers, of .025-inch Kalwall Sun-Lite, have a 1½-inch rise. The treads, of .020-inch shiny aluminum (or, in some instances, .025-inch Kalwall Sun-Lite to which a shiny aluminum sheet has been affixed), are 3 inches wide. In winter, much solar radiation passes through the staircase-like structure and enters the building proper: some radiation passes directly through the cover and the vertical Sun-Lite (the riser); some other radiation strikes the horizontal aluminum treads, is reflected and then enters via the risers, and subsequently encounters the underside of the next tread and is thence reflected downward into the building proper. The typical transmittance of the structure as a whole at noon on a sunny day in January is about 60 percent relative to the gross area of the roof, and about 70 percent relative to the net area of the staircase-like structure. Some radiation enters via the above-mentioned window. As regards heat loss by conduction and convection, the roof is said to be equivalent to 2 inches of fiberglass.

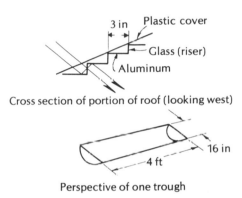

Cross section of portion of roof (looking west)

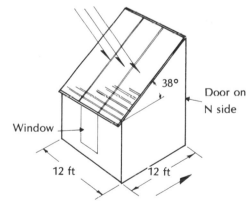

Perspective of one trough

Storage The special storage system includes twenty-seven troughs, in stepped array close beneath the roof, containing a total of 2½ tons of water. Each trough is 4 feet long and 16 inches wide and is oriented east-west. The depth at the center is 10 inches. The trough is made of .025-inch Kalwall Sun-Lite, with ends epoxy-cemented to the main, curved member. It holds 22 gallons (86 pounds) of water. A cover, of plastic film, prevents evaporation. The trough is supported, by its ends, from steeply sloping timbers that have a 12-by-4-inch cross section and are 4 feet apart on centers. These are the same timbers that support the roof structure. A small fraction of the solar radiation entering via the roof is absorbed by the troughs. A large fraction passes through them and either (1) penetrates deep into the room, heating and lighting it, or (2) encounters a system of lightweight vanes situated close below the troughs. Each vane is 4 feet by ½ foot, with the long axis running east-west. One face of the vane is aluminized; the other has a chrome green coating that absorbs the infrared portion of the solar radiation and much of the visual-range portion, but has a more pleasant appearance than black. When the room is hot enough,

Perspective view of bldg.

the occupant (or an automatic, thermally actuated device) turns the vanes so as to confine the solar radiation to the region occupied by the troughs. When the room becomes too cold, opening the vanes allows 4-to-40-micron radiation from the overhead troughs to warm the room (and, during a sunny day, allows much direct solar radiation to penetrate into the room).

Percent Solar-Heated   Near 100, if moderately large temperature swing is permitted.

Auxiliary Heat Source   None.

Domestic Hot Water   None.

Cooling in Summer   The solar roof excludes most of the solar radiation incident in summer. At noon, for example, the rays come from steeply overhead, and most are reflected (by the aluminum tread) back up toward the sky. At other times of day, an even larger fraction of the radiation incident is reflected toward the sky. Both because very little solar radiation enters the building and because such good insulation is used, little heat enters the building. Such heat as does enter produces only an extremely slow temperature rise because of the large thermal capacity of the water-filled troughs. The troughs are cooled late at night by circulation of outdoor air.

Note Concerning Illumination   The solar radiation entering the solar roof supplies not only heat but also light: a fraction of the radiation passes downward through the transparent troughs and fills the building as a whole with light.

Other Comments   The system works well without any use of electricity. It works well even if no manual adjustments are made—there are no shutters to close at night. There is no local overheating; for example, the south portion of the floor does not become especially hot. Nor is there any annoying glare there. The solar radiation travels directly to the near-ideal location of storage. No transport of heat by a flow of water or a flow of air is required. There is nothing that can freeze, except perhaps in the most unlikely circumstances. If any freezing did occur, it would do little or no damage. If there were damage, repair could be accomplished quickly and cheaply. The dampers are the only moving parts. There is no sound. The system is virtually vandalproof. The collection system is unaffected by the presence of 9-foot-high lilac bushes, etc., immediately to the south. To support the weight of water requires only simple strengthening of the building: the load which the water imposes is merely comparable to the nominal New England peak snow load. The presence of so much water close to hand may reduce fire risk.

Problems and Modifications   One significant drawback is that on some warm sunny days in winter the room temperature becomes uncomfortably high. In a violent storm in February 1978 some of the plastic roof covering, which had been installed relatively informally, was damaged.

Solar inventor, designer, supervisor: N. B. Saunders. *Builder:*   L. Tucker. *Owner:* Cambridge School.

System employing a two-area collector and a 100-ton bin-of-stones

Hopkins   43°N
(30 mi S of Grand
Rapids)

Building   This one-story toilet-and-shower facility for a private campground has a gross area of 2500 square feet and a heated area of 2000 square feet. There is no basement. The walls are protected by earth berms and 2 inches of Styrofoam. The ceiling is insulated with 9 inches of fiberglass. The window area is small, and the windows are double-glazed. The clerestory windows are glazed with Kalwall Sun-Lite. The building faces exactly south and has two south-facing roofs.

Collection   The air-type collector, with a gross area of 950 square feet and a net area of 800 square feet, is mounted on the two south roofs, which slope 53 degrees. It was fabricated on-site and is integral with the roofs. The absorber is a corrugated sheet of 29-gauge galvanized steel with a Rustoleum 412 nonselective black coating. The glazing is double and consists of Kalwall Premium Sun-Lite sheets ¾ inch apart. The inner sheet is ¾ inch from the black absorbing sheet. The airflow is in the 3½-inch space between the absorber sheet and the

> **Building:** 1-story, 2500 sq. ft.
> **Collector:** 950 sq. ft., air type
> **Storage:** 100-ton bin-of-stones
> **% Solar-heated:** See text

backing, which consists of 2 inches of rigid insulation. The air is driven upward within the collector at 3600 cfm by a 1-hp blower in the furnace room, in the north central part of the building.

Storage   The 100 tons (2100 cubic feet) of 2-inch-diameter stones is in a rectangular concrete-block bin in the south central part of the building. The bin is 25 by 12 feet by 7 feet high and is insulated on all sides and on the top; the top insulation consists of 9 inches of fiberglass. Hot air from the collector is driven downward through the bin. When the rooms need heat, room air is circulated upward through the bin by a ¾-hp blower associated with the furnace.

Percent Solar Heated   About 85 if the thermostat is kept at 55°F in midwinter.

Auxiliary Heat Source   Electric hot-air furnace.

Domestic Hot Water   This is preheated in a copper coil in the upper part of the bin.

Cooling in Summer   None.

Problems and Modifications   None of any significance.

Solar designer: Hawkweed Group, Ltd., and Khatib and Associates. *Architect:* Hawkweed Group Ltd. (esp. Rodney Wright). *Owner:* Sandy Pines. *Funding:* Private.

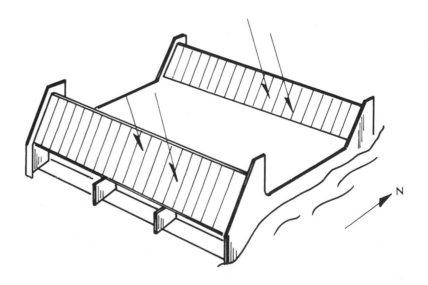

Federal Building

Free-standing collector 220 ft. long supports 8000 sq. ft. array of collector tubes

Saginaw   43½ °N
At Genesee,
Weadock, and
Warren Streets

Building   This reinforced concrete building, for use by thirteen Federal agencies, includes a post office, a loading dock area, and a parking area for 104 cars. Part of the horizontal roof is covered with earth in which grass, shrubs, and trees grow. Part is for car parking. Part serves as a playground. The building walls include large double-glazed window areas. Eaves with 10-foot outreach exclude direct radiation in summer. Many special energy-saving and water-saving features are used.

Collection   The collector is mounted on a tall, free-standing platform situated above the center of the building. The platform, which is 220 feet long and 44 feet high, is of precast concrete and slopes 45 degrees. The collector, with a total area of 8000 square feet, consists of vacuum-jacketed, water-type, tubular assemblies made by Owens-Illinois Incorporated. The coolant is water, with no antifreeze or inhibitor. It is circulated at 100 gpm by a 5-hp centrifugal pump and carries heat directly into the storage system, without intervention of a heat exchanger.

Building: 1-story, 59,500 sq. ft.
Collector: 8000 sq. ft., tubular type
Storage: 30,000 gals of water
% Solar-heated: 40 (predicted)

Storage   The 30,000 gallons of water is in two 15,000-gallon tanks. Heat is delivered to the rooms by perimeter baseboard radiators and also by fan-coil systems.

Auxiliary Heat Source   Oil-fired boiler.

Domestic Hot Water   This is heated by the above-described solar-heating system.

Cooling in Summer   An absorption-type chiller is used and is powered by 210°F water from the solar heating system or by water from the oil-fired boiler. It is expected that the solar heating system will supply 55 percent of the power needed.

Problems and Modifications   Since it started operating (on July 13, 1977), the solar heating system has performed without significant problems.

Final design: U.S.General Services Administration, through Public Buildings Services, Construction and Management Division, Chicago, Illinois; Project Manager: R. D. McGinnis. (An earlier, superseded design was prepared by Smith, Hinchman and Grylls, Associates, Incorporated.) *Fabricator of tubular collector assemblies:* Owens-Illinois Incorporated. *Builder of the solar collection system:*   R.C.Hendrick and Sons, Incorporated. *Coordinator of design of solar collection system:*   T. J. McNamara. *Cost of entire project:*   About $4,600,000. *Cost of collector and associated piping, etc.:* $272,000 (estimated). *Funding:*   Mainly by General Services Administration, partly by Energy Research and Development Administration. *Instrumentation:*   By University of Michigan, under contract with the General Services Administration and with funding by Energy Research and Development Administration.

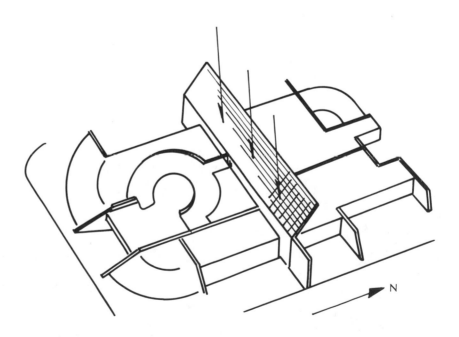

Passive system employing two sets of south-facing windows provides 90 % solar heating

Building This three-bedroom house is of post-and-beam construction with brick veneer and a clerestory that runs the length of the house. In addition to the bedrooms the house includes a kitchen-dining area, a den, three bathrooms, a foyer, and a workroom; also a mezzanine and bridge over the den. There is an attached garage. The walls and roofs are insulated to approximately R-30, and many precautions have been taken to minimize leakage of air. The windows are double-glazed; at night they are covered by heavy curtains. The covering of the south roof consists of .032-inch aluminum.

Passive Solar Heating System The clerestory window array is 54 feet long by 5 feet high. Solar radiation passing through these windows penetrates deep into the north part of the house, warming it. The south windows of the first story also admit much radiation to the south part of the house. The total area of south-facing windows is 350 square feet. The aluminum-covered south roof reflects much radia-

Longview   33½ °N
(a 2800-degree-day site)
(30 mi W of Columbus)
On Highway 12

Building: 1½-story, 2600 sq. ft.
Collector: } Passive
Storage: }
% Solar-heated: 90

135

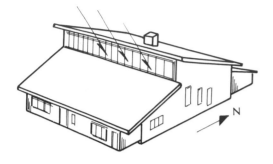

tion toward the clerestory windows. Thermal storage is provided by the mass of the house itself.

Auxiliary Heat Source   Circulator-type fireplace and one portable 1-kw electric heater.

Domestic Hot Water   This is not solar-heated.

Cooling in Summer   One 18,000-Btu/hr conventional window-type air-conditioner is used. Little cooling is needed because of the insulation used, the wide eaves that, in summer, intercept radiation approaching the first-story windows and clerestory windows, and the aluminum-covered south roof that reflects incident radiation toward the sky.

Solar engineer, general engineer, designer, builder: Pablo Okhuysen. *Consulting engineer:*   J. L. Ferguson; also W. L. Croft. *Owner and occupant:*   W. L. Croft.

Thomsen Retrofit House

A 670-sq.-ft. air-type collector, with 10-ton bin-of-stones, provides this very prevalent type of Nebraska house with 60 % solar heating

Lincoln   41°N
1141 Carlos Drive

Building   A solar heating system was installed in an existing one-story, 1500-square-foot, three-bedroom wood-frame house with full basement, attached garage, and no attic. Walls and roof are insulated to R-11 and R-30, respectively. The windows are double-glazed. Apart from its solar heating system, this house is typical of hundreds in the neighborhood.

Collection   The collector, which consists of a single assembly fabricated on-site, is 56 feet long and 12 feet wide and is on a roof sloping 12 degrees. The absorber is a sheet of V-corrugated .032-inch aluminum ("V-beam roofing") that has a nonselective black coating and is oriented with the corrugations running east-west. The glazing is single and consists of corrugated Filon (polyester and fiberglass with a Tedlar coating) oriented with the corrugations north-south. The collector backing includes 1½ inches of semi-rigid fiberglass. In the space behind the absorber sheet, air flows west along certain portions of the collector and east along other portions; thus the supply and re-

> **Building:** 1-story, 1500 sq. ft.
> **Collector:** 670 sq. ft., air type
> **Storage:** 10 tons of broken brick
> **% Solar-heated:** 60 (predicted)

turn header ducts are at the same end. A flowrate of 900 cfm is maintained by a ½-hp blower. The ducts are 1 foot in diameter, and the pressure drop in the collector is 1 inch of water. The collector is pierced by furnace chimney and plumbing vents; flashing and caulking prevent leakage of air or rain.

Storage   Heat is stored in 10 tons of broken bricks in a rectangular wood-frame bin in the furnace room. The bin, 9 by 5 feet by 7 feet high, is lined with plywood and is insulated with fiberglass batts covered with gypsum wallboard. Hot air from the collector passes downward through the bin. Heat is distributed to the rooms by a duct-and-blower system that normally serves the furnace. The control system includes five motorized dampers. The pressure drop in the bin is ⅓ inch of water.

Auxiliary Heat Source   Existing gas-fired furnace and hot-air distribution system.

Domestic Hot Water   This is preheated in a small metal tank within the central upper region of the bin. Final heating is by a gas-fired heater.

Cooling in Summer   A conventional air conditioner is used.

Problems and Modifications   Initially, the walls and ceiling were insulated to R-4 and R-6, respectively. This was found to be insufficient and, in 1977, the insulation was increased to R-11 and R-30, respectively. Late in 1977 a small greenhouse was being constructed, attached to the south side of the house.

Solar engineer, designer, owner, occupant: C. L. Thomsen of the firm Clark Enerson Partners. *Funding:*   Private. *Cost of solar heating system:*   Materials, $2500; labor; $2500.

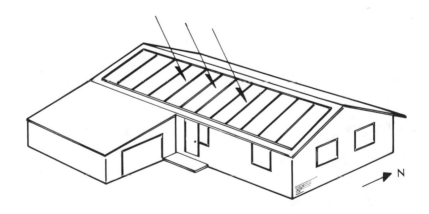

Freese House

Water-filled plastic bags in attic receive solar radiation via sloping Beadwall roof

Concord   43°N
Fisk Road

Building   A new wing was added to an existing farmhouse, and a solar heating system, serving just the wing, was installed early in 1976. The wing, which is 34 feet long and 20 feet wide, has two stories, an attic, and crawl space, but no basement. The heated area is 1360 square feet. The first story contains the living room, kitchen, and laundry, and the second story has one bedroom, a large study, and a bathroom. The walls are insulated with 6 inches of fiberglass and the roof contains 12 inches. The foundation walls, of granite, are insulated on the inner faces with 2 inches of urethane foam. The windows are double-glazed.

Solar Heating System   The solar heating system is largely passive. Solar radiation enters the attic via the south face of the roof, which slopes 60 degrees and is double-glazed with Kalwall Sun-Lite. The glazed area (288 square feet) is 32 feet long and 9 feet high. The two glazing sheets are 4 inches apart and are part of a Beadwall system. On sunny days in winter there is only air between the two sheets, and radiation passes through freely and strikes a large horizontal array of thirteen water-filled black vinyl bags resting on the attic floor. The

Building:	2-story addition, 1360 sq. ft.
Collector:	See text
Storage:	
% Solar-heated:	See text

total amount of water is 2000 gallons, or 300 cubic feet, weighing 9 tons. Each of the bags is approximately rectangular and is 7 by 6 feet by 10 inches high. Warm air from the space above the bags is sent, by a blower, downward into the crawl space beneath the first-story floor and thence upward into the first-floor rooms via five vents. The return air flows up the circular stairwell and into the attic. Some attic air is sent downward directly to the second-story study and, indirectly, to the bathroom. At the end of a sunny day a small blower blows a large quantity of 3/32-inch-diameter spheres of polystyrene foam into the space between the two Sun-Lite sheets, to provide insulation equivalent to about 4 inches of fiberglass. The above-described solar heating system, which is mainly passive but partly active, is supplemented by purely passive solar heating provided by the large vertical window areas on the south face of the building.

Percent Solar-Heated   No estimate is available. Guesses range from 30 to 60 percent.

Auxiliary Heat Source   There is an oil furnace with four-zone hot-water distribution system. Prestone is included in the water so that even if the house were left completely unheated for a long period in winter, no freeze-up would occur.

Domestic Hot Water   This is not solar-heated.

Cooling in Summer   Operation of the Beadwall roof and the attic storage system is reversed, to provide cooling. The Beadwall remains filled with beads, to exclude solar radiation, and during the night, cool outdoor air is circulated through the attic by means of a blower in order to cool the water-filled bags. Cool air adjacent to the bags is circulated, via the crawl space, to the rooms.

Solar consultant: Bruce Anderson. *Architect:* E-M Architects (especially B. C. Ellis). *Builder:* Community Builders (esp. Don Booth). *Owner and occupant:* Donald Freese. *Funding:*   Private.

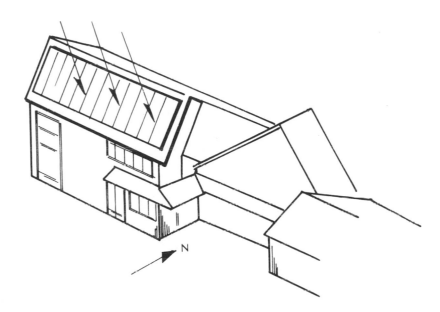

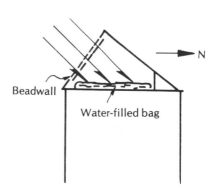

Partial cross section looking west

Grantham   43½°N
(near Lebanon)
In Eastman
Development

Building: Split-level, 1530 sq. ft.
Collector: 700 sq. ft., air type
Storage: 1500 gals. water
% Solar-heated: See text

Air-type system employing bin filled with 1500 water-filled polyethylene bottles

Building   This is a split-level, three-bedroom wood-frame building in a small clearing in the woods. The basement includes one bedroom, a combination utility room and darkroom, and the storage system. The house has a small attic space. There is no garage. The window area is small, and there are no windows on the south side of the house. All windows are of Anderson casement Thermopane type. The walls are insulated with 3½ inches of fiberglass and 1 inch of foamed-in-place urethane. The roof is insulated with 8 inches of fiberglass and 1 inch of foamed-in-place urethane. The estimated heat loss of the building is 15,000 Btu per degree-day.

Collection   The air-type collector, on the 60-degree-sloping south face of the house, is 700 square feet in gross area and 640 square feet in net area. The absorber is a sheet of .020-inch aluminum with a nonselective black coating. The glazing is double, consisting of two sheets of Kalwall Sun-Lite Premium ½ inch apart. The outer sheet is .040 inch thick and the inner sheet is .025 inch thick. The collector backing includes 1 inch of urethane foam. Air is driven upward in the space behind the absorber sheet at 3000 cfm by a ⅓-hp blower in the basement. Along the bottom and top of the collector there are header manifolds 1½ by 1½ feet in cross section.

Storage   The 1500 gallons of water is contained in 1500 one-gallon polyethylene bottles arranged in spaced manner in a rectangular bin in the north part of the basement. The bin is 32 by 12 feet by 6½ feet

high. The top of the bin is insulated with 6 inches of fiberglass, and the sides are insulated with 1 inch of urethane foam. Air from the collector flows upward through the bin. Usually, the temperature of the bottles does not exceed 95°F. If the bottles are insufficiently hot, a General Electric Company 3½-ton air-to-air heat-pump is turned on automatically. In winter its sole function is to extract heat from the bin and deliver the heat to the rooms.

Percent Solar-Heated   Solar heating provides about 55 percent of the winter's heat need. The combination of the solar-heating system and the heat-pump provides about 70 percent.

Auxiliary Heat Source   15 kw electric heater, operated at off-peak times to heat the storage system.

Domestic Hot Water   This is preheated in a small tank within the bin-of-bottles.

Cooling in Summer   The heat-pump, operated in reverse manner, cools the rooms directly. Also, the bin-of-bottles is cooled at night by cool outdoor air circulated through it, and during the day, the bin can be used to cool the rooms.

Problems and Modifications   Some of the Kalwall Sun-Lite sheets were not secured properly. The situation was soon corrected. Initially, each row of water-filled polyethylene bottles rested on a sheet of plywood resting on the next row below; when the bottles became hotter than 80°F they showed signs of being compressed vertically. Accordingly, cinder blocks were installed between plywood sheets to take the weight.

Solar engineer: Dick Gregor. *Contractor:*   G. Collier. *Owner, occupant:*   A. Rieger. *Funding:*   Private. *Cost of solar heating system:*   About $4500 additional, relative to system using heat-pump and electric heaters.

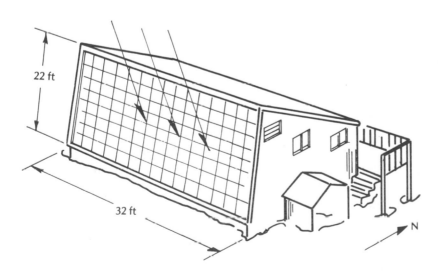

Goosebrook House

Harrisville   43°N
(15 mi NE of Keene)

Building: 2-story, 1400 sq. ft.
Collector: ⎱
Storage: ⎰ See text
% **Solar-heated:** 60 (predicted)

House with unique combination of active and passive collection, storage, and heat-conservation systems

Building   This three-bedroom, wood-frame house has a full basement, a small attic space, an attached greenhouse at the southeast corner, and, at the southwest corner, an attached two-car garage. The walls are insulated with 5½ inches of fiberglass and 1 inch of tongue-and-groove boards of polystyrene foam. The attic ceiling is insulated with 15 inches of cellulose loose fill. The basement walls are insulated externally with 3 inches of polystyrene foam. The windows on the south side are large and are an element of the passive system described below. The windows on the east, north, and west sides are small and are triple-glazed. The garage, which has a loft, is not insulated or heated. The house faces exactly south.

Solar Heating System   There are three parts to the solar heating system: a passive system employing large vertical south windows, an active system employing a water-type collector on the sloping south roof, and a greenhouse.

Passive System. Radiation is received via one large (110-square-foot) vertical south window-wall, 16 feet long and 7 feet high. The windows are double-glazed with Thermopane. They are partially shaded in the early morning by the greenhouse and in the late afternoon by the garage. Some of the passively received energy is stored in nine vertical, water-filled cylinders that stand on the living room floor 1 foot from the big south window-wall. The cylinders, made of transparent Kalwall Sun-Lite, are 12 inches in diameter and 8 feet high. Each contains 380 pounds of water, and the set of nine contains 3400 pounds. At night the window-wall is covered with two 3-inch-thick sliding shutters, each 8 feet long and 7 feet high, faced on front and back with ¾-inch-thick boards, with a 1½-inch plate of polystyrene foam sandwiched between. Each shutter weighs 400 pounds. The two shutters are in tandem and are suspended from (and travel along) the same overhead east-west track, of the type used for a New England barn door, which extends into the garage. Each shutter is suspended by hangers provided with ball bearings. When not in use, the shutters are far to the west, inside the garage, between the two car-parking spaces. When the shutters are to be closed, they are manually pushed eastward, one at a time, by a person standing in the garage. When they are closed, covering the window-wall, sealing strips along all four edges provide tight closure. A guide-fin projecting downward from the bottom of each shutter engages an east-west slot in the porch and prevents the lower part of the shutter from swinging outward away from the window-wall. (The likelihood that wind will affect the shutters is small, inasmuch as the shutters are deep within a pocket defined by the greenhouse, the garage, and the 4-foot roof overhang.)

Active System. There is a 480-square-foot trickling-water-type collector on the roof, which slopes 59 degrees. The collector is 28 feet long and 17½ feet high. The collector panels were made on-site, following a design somewhat similar to that pioneered by Thomason Solar Homes, Incorporated. The absorber is a blackened corrugated sheet of aluminum, with the corrugations running up and down the roof. The glazing is single, consisting of double-strength glass panes 4 by 3 feet. The collector backing, which is the roof itself, includes 6 inches of fiberglass. The coolant is water, with no antifreeze; the water is drained automatically whenever the supply pump stops running. Near the ridge of the roof there is a near-horizontal supply pipe that delivers water to each valley of the corrugated sheet. At the bottom the water is collected by a gutter. Water is supplied to the collec-

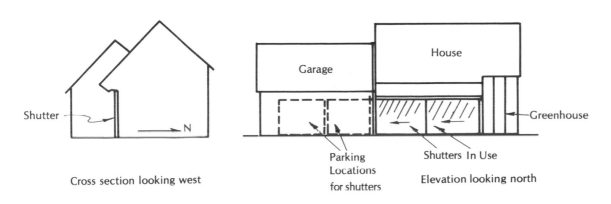

Shutter

N

Cross section looking west

Garage

House

Parking
Locations
for shutters

Shutters In Use

Greenhouse

Elevation looking north

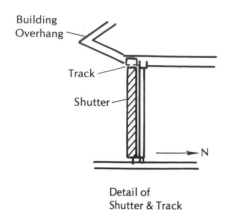

Detail of
Shutter & Track

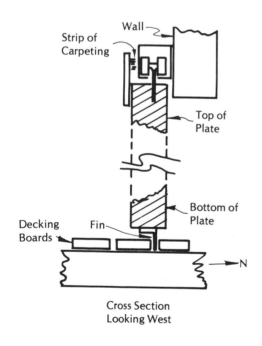

Cross Section
Looking West

tor at 7 gpm by a ½-hp centrifugal pump. The associated storage system employs 14,000 pounds of water stored in eighteen vertical, cylindrical tanks 18 inches in diameter and 7½ feet high. Each tank is made of Kalwall Sun-Lite. The tanks are arranged in two east-west rows of nine tanks each. Within each row, the tanks are 21 inches apart on centers; thus there are 3-inch airspaces between tanks. The tanks are housed within a slender, rectangular, insulated bin in the north part of the basement. The bin is 18 feet long and 4 feet wide, with the long axis running east-west. The eighteen tanks are connected so as to form three sets of six each. Within each set the tanks are in series, the water circulating from the top of one tank to the bottom of the next; the flow has an east-to-west trend, so that the eastmost tank is usually the hottest. The three sets are hydraulically in parallel; each receives hot water directly from the collector. When the rooms need heat, a ¼-hp fan circulates room air eastward through the bin, traveling first past the cooler tanks and finally past the hottest tanks. It is expected that the amount of heat provided by the active system will be much greater that that provided by the passive system.

Greenhouse. The greenhouse is 10 by 8 feet, and its single-glazed 10-by-10-foot south window slopes 59 degrees. When the greenhouse is warm and the rooms are cold, the door between greenhouse and house proper may be opened to allow some of the warm air to circulate into the rooms.

Auxiliary Heat Source A Megatherm electric heater with a 210-gallon tank. It is operated at off-peak hours. The temperature of the water, which is pressurized, can reach 280°F.

Domestic Hot Water This is preheated in a copper-tube coil immersed in the upper part of one of the hottest of the vertical tanks in the storage bin.

Cooling in Summer Little cooling is needed, inasmuch as (1) the building is very well insulated, (2) the roof overhang shields the window-wall in summer, (3) the greenhouse can be isolated from the house proper, and (4) the thermal capacity of the living room is greatly increased by the row of slender water-filled tanks. On especially hot and sunny days the occupants can close the shutters serving the window-wall.

Problems and Modifications During the completion of construction and preliminary operation, several difficulties were encountered, including inadequate capacity of the gutter; imperfect confinement of moisture condensing on the underside of the collector glazing; incorrect size of water-pump, with consequent cavitation and excessive noise; and inadequate capacity of the manifold connections among the vertical tanks in the storage bin. Most of the difficulties were soon overcome.

Solar engineer and architect: Total Environmental Action (esp. Bruce Anderson and Charles Michal). *Builder:* George Bogosian and Company. *Owner and occupant:* No information. *Cost of house, land, and solar heating system:* About $65,000 to $75,000. *Cost of solar heating system:* About $8500. *HUD grant:* $8,579.

Two air-type collectors and 50-ton bin-of-stones provide 50 to 60% solar heating

Lyme 44°N
(50 mi NW of Concord)

Building This four-bedroom house, 36 feet long and 27 feet wide, has a full basement, entry porch, viewing deck, small attached greenhouse, and attached one-car garage. The north side of the basement has an on-grade walkout. The living room and dining room have cathedral ceilings. The walls are insulated to R-21 and the ceilings to R-40. The concrete foundation walls are insulated on the inside to R-19. The total area of view windows is 315 square feet. All of the windows are double-glazed. Many precautions have been taken to minimize air leakage.

Building: 2-story, 2000 sq. ft.
Collector: 600 sq. ft., air type
Storage: 50-ton bin-of-stones
% Solar-heated: 50 to 60 (predicted)

Collection The air-type collector, mounted on a vertical face of the house proper and on a vertical face of the garage, has a gross area of 600 square feet and a net area of 535 square feet. The collector panels, built on-site, employ ribbed sheets of .040-inch aluminum with a nonselective black coating. The glazing is double, consisting of two sheets of Kalwall Sun-Lite. Blower-driven air travels in the space behind the aluminum sheet.

Storage   The 50 tons of 1½-inch-diameter stones (crushed rock) is in a rectangular wooden bin insulated to R-40. (Note: For further details as to the collector and storage system, see the account of the Wilder House in Hartford, Vermont, page 242.)

Auxiliary Heat Source   Oil-fired boiler with water-to-air heat exchanger. Also a 45,000-Btu/hr cast-iron wood-burning stove with air supplied directly from outdoors.

Domestic Hot Water   This is not solar-heated.

Cooling in Summer   None.

Solar engineer: Total Environmental Action, Incorporated. *Designer and builder:* CHI Housing, Incorporated. *Owner, Occupant:* R. Wickware. *Funding:* Private.

Environmental Education Center

Water-type system installed in large public building

Building   This wood-frame structure includes a main building, two wings, and a full basement with ground-level entrances. There is a small attic space and an observation deck. The walls are insulated with 4 inches of fiberglass and the ceilings or roofs contain 6 inches. One quarter of the 1500 square feet of window area is on the south side of the building. All windows are double-glazed. The building faces 14 degrees east of south.

Collection   The collector is incorporated in the main roof, 58 by 35 feet, and the west-wing roof, 90 by 34 feet. Each roof slopes 18½ degrees. The 135 collector panels were made by General Electric Company. Each is 8 by 3 feet and slopes 40 degrees—considerably more steeply than the roof itself. The roof is single-glazed with ¼-inch Lexan sheets supported by aluminum framing. The collector panels, also single-glazed with Lexan, are suspended close beneath the roof: Thus, in effect, the panels are double-glazed (with sheets sloping 18½ and 40 degrees). The absorber sheets in the panels are of Olin Brass Company Roll-Bond aluminum with integral passages for coolant. The selective black coating has an a/e ratio of .94/.34. The coolant is a 50-50 solution of water and Prestone. To minimize the danger of corroding the aluminum, the designers called for (a) using deion-

Basking Ridge   40½°N
(30 mi W of New York City)
190 Lord Stirling Road, at Lord Stirling Park

Building: 1-story, 18,000 sq. ft.
Collector: 3100 sq. ft., water type
Storage: 6000 gals. water
% Solar-heated: 50 to 60

ized water, (b) using no copper or brass fittings—the valves, pipes, etc., are of iron or steel, (c) separating the aluminum sheets from the iron and steel hardware by means of plastic couplings, (d) installing a 50-micron filter to prevent small particles of iron, steel, etc., from reaching the aluminum sheets. The backing of the panels includes 2 inches of isocyanurate foam. The coolant is circulated at 65 gpm by a 5-hp centrifugal pump. Heat is transfered to the storage system by a heat exchanger.

Storage   The 6000 gallons of water is in two tanks of 4000-gallon and 2000-gallon capacity. Each is a horizontal, cylindrical, 5½-foot-diameter steel tank, insulated to R-30 with 4 inches of Owens-Corning urethane foam #240 covered with Cadalar 670 vapor barrier. The tanks are situated outside the building, adjacent to the east side; they are 3 to 4 feet underground and rest on well-drained beds of sand. The rooms are heated by a forced-water system serving thirty-four fan-coil units.

Auxiliary Heat Source   A Kewanee KP-5 oil-fired boiler. When auxiliary heat is needed, a Honeywell Controller adjusts the proportion of water heated by the boiler to water (from the storage system) not heated by the boiler so that the temperature of the water fed to the fan-coil units is the minimum compatible with keeping the rooms warm. For example, if the outdoor temperature is 40°F, water at 90°F is hot enough. Thus the amount of heat extracted from the storage system is maximized, the temperature of the storage system is relatively low, and accordingly collection efficiency is high.

Domestic Hot Water   This is preheated by the solar heating system. Final heating is by the above-mentioned oil-fired boiler.

Cooling in Summer   An Arkla Servel Model WF-300 absorption chiller, employing lithium bromide and rated at 300,000 Btu/hr, is used. On sunny days it is powered by hot water from the solar heating system collector or the storage system, with backup by the oil-fired boiler mentioned above. Or a conventional Webster CP30A chiller can be used. Note: Only half the floor space, i.e., 9000 square feet, is cooled.

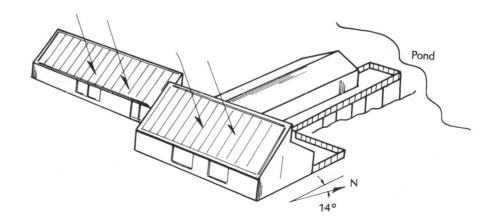

Problems and Modifications   The original contracting company that was to construct the building went bankrupt and the process of locating another contractor resulted in a delay of about a year. Partly because of this delay, warranties on many components of the solar heating system expired before the start-up of operation; but, fortunately, there were very few failures. Two minor leaks in the collector system occurred as a result of incorrect tightening of two fittings. Because each portion of the collector system was pressure-tested separ-

ately during installation, and likewise each assembly of portions, no significant leaks occurred during routine operation. Leaks from the collector would be especially unwelcome inasmuch as the collectors are situated indoors, close beneath roof, and in some areas any water-and-Prestone that leaked could drip onto people and furnishings. One differential controller was found to be defective, and one modulating valve. The insulation on the undersides of the collector panels (2-inch, foil-faced, high-temperature, isocyanurate insulation) had too low an R-value: during the hottest days in summer, with collector temperature sometimes reaching 235°F, radiation from the panels delivered so much energy to the living space below that the air conditioning system could not keep the space comfortable. Because the city electrical supply sometimes fails and because such failure could interrupt the flow of coolant to the panels, with resultant danger of deposition of contaminants and consequent corrosion, the provision of a small standby gasoline-powered generator was deemed necessary. Also, because fluctuations in voltage or frequency of the city electric supply sometimes occur and could damage data-logging equipment, the provision of voltage and phase regulation was deemed necessary. (One data-logger was damaged by such fluctuation.) The nonunion supplier of a major instrumentation console was prohibited by state law from making the installation; introducing a union contractor to make the installation entailed delay and extra cost. The overall cost overrun on the solar heating and cooling system was of the order of $300,000.

Solar heating system design: Becht Engineering Company. *Manufacturer of the solar collector:* General Electric Company. *Architect:* Halsey and Ryder. *General planning:* Environmetal Education Center of Somerset County Park Commission, W. Jones, Director. *Solar energy coordinator:* P. C. Becker. *Principal investigator:* J. W. Moody of Somerset County Park Comission. *Funding of building proper:* Somerset County Park Commission. *Funding of solar heating and cooling system:* Energy Research and Development Administration, by contract providing $459,000 in first year. *Funding of evaluation studies:* $97,000 by Energy Research and Development Administration. *Total final cost of solar heating and cooling system:* $965,000. (Note: This figure includes the costs of several supplementary items, e.g., electronic systems for recording and analyzing performance data, educational displays, and a half-hour motion picture explaining the project as a whole.)

Kelbaugh House

House with large-area, thick, two-story Trombe wall and a small attached greenhouse

Princeton   40°N
9 Pine Street

Building This is a wood-frame house, except for the south side which is of glass and concrete. Attached to the southeast corner of the house is a small greenhouse. Below it there is a small, heated basement. The house is well insulated, the R-values of typical walls and roof being 17 and 33 respectively. The house windows are triple-glazed and the greenhouse windows are double-glazed. The stairwell is closed by a door to prevent flow of warm air from first story to second story.

> **Building:** 2-story, 2100 sq. ft.
> **Collector:** }
> **Storage:** } Passive
> **% Solar-heated:** 75 to 80

Passive Solar Heating System Radiation enters the entire vertical south face, which is double-glazed. Spaced 6 inches from this window-wall is a vertical, 15-inch-thick concrete wall, the south side of which has a somewhat selective black coating. Every room except the bathrooms has this concrete wall as its south wall, and in each of these rooms cooler air near floor level passes, via rectangular ports in the concrete wall, into the 6-inch airspace between wall and glazing. Here the air is warmed, rises, and passes back into the room via ports near ceiling height. Circulation is by gravity convection. Some heat is

supplied to the rooms directly by conduction of heat through the wall. Reverse gravity-convective flow of air through the ports at night or on overcast days is prevented by automatic one-way check valves, or dampers, consisting of an extremely thin (about .001-inch) vertical sheet of polyethylene attached along its upper edge only and a fixed vertical screen. During a sunny day the direction of airflow is such as to push this sheet to open position, but at night the flow tends to reverse, pushing the sheet against the screen and blocking all flow. Storage is provided by the massive south wall, the concrete floor of the greenhouse, and by eight water-filled, black 55-gallon steel barrels in the greenhouse. The solar heating system provides 75 to 80 percent solar-heating if the rooms are kept, on the average, at 65°F and if temperature excursions five degrees above or below are accepted.

Auxiliary Heat Source   Hot-air furnace. Ducts carry hot air to all main rooms and the bathrooms.

Domestic Hot Water   This is not solar-heated.

Cooling in Summer   The space between the glazing and the massive wall is well vented. When natural venting is insufficient, four fans serving four vents along south edge of roof are turned on. As hot air flows out of the house via these vents, cool air flows in via windows on the north side of the house. The greenhouse is covered by an external roll-down shade.

Problems and Modifications   Initially an excessive amount of warm first-story air flowed up the stairwell to the second story. The difficulty was largely overcome by installation of a stairway door. Some reverse thermosiphoning of air between south wall glazing and concrete wall occurred at night until the installation of the check valves described above. The greenhouse was sometimes too cold at night; accordingly, a second layer of glazing was added. The greenhouse sometimes became too hot during sunny days; installation of water-filled barrels reduced such temperature rise. On cold nights in winter much heat is lost through the large double-glazed south wall. It is believed that the cost of installing operable shutters to reduce such loss would be excessive.

Solar engineer, architect, owner, occupant: Doug Kelbaugh.

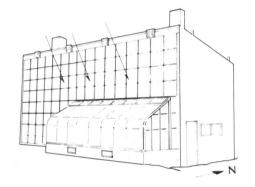

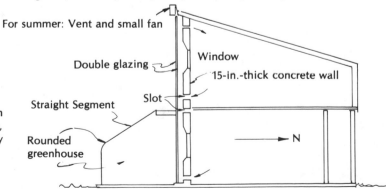

Note: Omitted from this diagram are: basement under greenhouse, footings of building and many other details

For summer: Vent and small fan

Double glazing

Window

15-in.-thick concrete wall

Slot

Straight Segment

Rounded greenhouse

N

**Partly underground house, with earth-and-plant-covered roof, 75%
solar-heated by trickling-water-type collector**

Shamong   40°N
Township
(in Burlington
County)
On Tuckerton Road,
near Route 206 in
Indian Mills

Building   This is a three-bedroom house 100 feet long and 24 feet
wide. It has a very small basement and no attic. The south roof,
which includes the collector and a very small window area, slopes 70
degrees. The north roof, which slopes oppositely 20 degrees, is strong
and well-insulated and is covered with 1 foot of mulch and 1 foot of
earth; grass and shrubs grow in the earth. The window area on the
south side of the house, below the collector, is large: 290 square feet.
The window areas on the other sides are very small. All windows are
double-glazed. The house faces 5 degrees west of south.

Collection   The collector, sloping 70 degrees, is of Thomason trickling-water type and is 100 feet long, 7 feet high. It includes 24 panels,
each 7 by 4 feet, obtained from Edmund Scientific Company. The
black coating on the corrugated sheet of aluminum is nonselective.
Each panel has a pipe at the top to distribute water to each valley of
the corrugated sheet and a collector-pipe at the bottom. The water,

---

Building: 1-story, 2400 sq. ft.
Collector: 670 sq. ft., trickling-water type
Storage: 2000 gals. water and 30-ton bin-of-
           stones
% Solar-heated: 75

which contains no antifreeze, is circulated by a ⅓-hp pump. The panels are single-glazed with 1/8-inch glass sheets 32 by 24 inches.

Storage   The 2000 gallons of water is contained in a horizontal, cylindrical steel tank 5 feet in diameter and 12 feet long. The tank has an access manhole at the top. Surrounding the tank is a 30-ton quantity of stones 4 to 7 inches in diameter. The stones are in a rectangular concrete-block bin, the dimensions of which are 20 by 10 feet by 8 feet high. The bin is near the center of the house, below the dining room. The tank, containing hot water from the collector, warms the stones, which in turn warm air (driven by two ½-hp blowers) that is circulated to the rooms via large ducts.

Auxiliary Heat Source Two fireplaces and an oil-fired hot water heater. This latter, situated in the utility room just north of the bin-of-stones, contributes heat to the bin via a flue-pipe loop (used in winter only) and via a set of finned tubes fed water from the hot-water heater. The flue-pipe loop and the finned tubes are in the upper part of the bin.

Domestic Hot Water This is preheated in a 40-gallon galvanized steel tank within the above-mentioned 2000-gallon tank.

Cooling in Summer The big tank and the bin-of-stones are cooled at night by a conventional 2½-ton air conditioner, which discharges heat to the cool night air. During subsequent hot days, room air is circulated through the cool bin.

Problems and Modifications Initially, the water circulated to the collector was not filtered, with the result that some discoloration of the black absorbing surface occurred. A filter was soon installed.

Solar engineer: H. E. Thomason. *Architect:* M. B. Wells. *Owner and general contractor:* Robert Homan. *General assistance:* Edmund Scientific Company. *Cost of building:* About $60,000; the cost might have been 25 percent higher had not various components been donated by assisting groups. *Incremental cost of solar heating system, relative to conventional system:* $6500 (estimated).

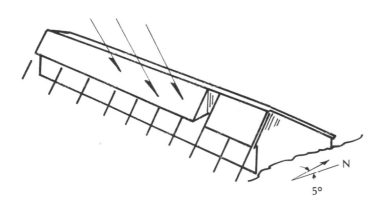

N

5°

Bridgers-Paxton Building

Albuquerque   35°N
213 Truman Street,
NE

Building: see text
Collector: 750 sq. ft., water type
Storage: 6000 gals. water
% Solar-heated: see text

Office building heated by the combination of a solar collector, a storage system, and five heat-pumps

Period 1956-1962

Building   Before 1962, this one-story, well-insulated office building had a useful floor area of 4300 square feet. The building was in two parts (north and south) and the collector was installed on the south face of the north part. The drafting room ceiling, behind the collector, was 14 feet above floor level.

Collection   The 750-square-foot, water-type collector consisted of fifty-five panels in one large array sloping 60 degrees. Twenty-six of the panels were 9 feet by 22 inches, twenty-six were 7 feet by 22 inches, and three were 5 feet by 22 inches. Each panel included, as absorber, a 3/16-inch-thick aluminum sheet, to the back of which ½-inch-diameter copper tubes, 6 inches apart, were soldered. The black paint used was nonselective. The glazing was single, consisting of double-strength glass. There was a 4-inch airspace between the absorber and the glass. The coolant was water; no antifreeze was used, and the water was drained before freeze-up could occur.

Storage   The 6000 gallons of water was stored in a cylindrical, 5-foot-diameter steel tank (not insulated) buried in the ground beside the building. When the water was insufficiently hot, an Acme 7½-ton water-to-water heat-pump (water chiller) was used to transfer energy from the tank water to the water that was circulated to the room heating system. When the input and output temperatures were 55° and 105°F respectively, the heat-pump had a coefficient of performance of about 4.5.

Auxiliary Heat Source   None.

Cooling in Summer   The heat-pump was used in cooling mode.

Period 1962-1974

The building was enlarged, the floor area being doubled. A conventional furnace was installed and put into operation. The solar heating system was left idle.

Period after 1974

The solar heating system was put back into use, with some modifications. As coolant, water and ethylene glycol was used instead of water alone, and a heat exchanger was installed between collector and storage system. Also, five small water-to-air heat-pumps were installed to extract heat from the storage system and deliver the heat to five groups of rooms. Four of the pumps were made by American Standard and one by American Air Filter Company. Together they consume 10 kw. Typical coefficients of performance are 3 to 3.5. About 60 percent of the building is 100 percent heated by the combination of the solar heating system and the heat-pumps. Operation has been trouble-free.

Solar engineers, owners: F. H. Bridgers and D. D. Paxton. *Architect:* Stanley and Wright. *Performance study:* By S. F. Gilman of Pennsylvania State University under 1975 grant from National Science Foundation.

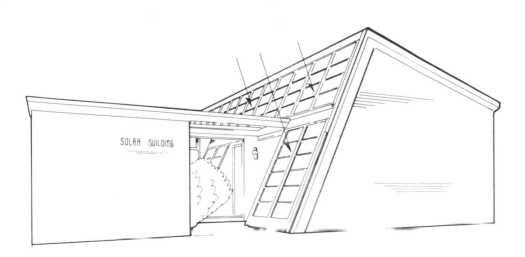

Kusianovich House

Albuquerque   35°N
2324 Morrow Road,
NE

Building: 1-story and heated basement,
2880 sq. ft.
Collector: 270 sq. ft., concentrating type
Storage: 2000 gals. water
% Solar-heated: 80

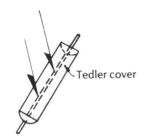

Collector box

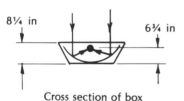

Cross section of box

A rooftop array of 20 cylindrical, parabolic collector boxes provides 80% solar heating

Building   This house, with one main story and a full, heated basement, is 60 feet long and 24 feet wide. The heated floor area, including basement, is 2880 square feet. At the northeast corner of the building there is an attached 2-car carport. There is a greenhouse within the southeast portion of the building. The total window area on the south side of the building is 140 square feet. The window areas on the other sides are much smaller. All of the windows are double-glazed. The roof, which slopes 14 degrees, is insulated with 6 inches of rockwool.

Collection   The 270-square-foot collector, which slopes 50 degrees, is mounted on a special stand on the 14-degree-sloping south roof. The collector consists of a row of twenty Solcan collector boxes, made by Albuquerque Western Solar Industries, Incorporated. Each box is 96 inches long, 24 inches wide, and 9 inches deep and is oriented with its long axis up to the north, down to the south, at 50 degrees to the horizontal. The boxes are 30 inches apart on centers and form a single row. In the upper part of the box there is an 8-foot-long, 1.25-inch-diameter Type M copper tube that has a nonselective black coating. Below the tube, and parallel to it, there is a reflector of aluminized .005-inch mylar with the aluminized face up. This reflecting film is held in near-cylindrical-parabolic shape by four transverse ribs of spring steel. The boxlike housing is made of angle irons and galvanized steel sheets and the glazing (window) consists of a single sheet of .004-inch Tedlar. The coolant is water and corrosion inhibitor. There is no antifreeze; the coolant drains automatically before freeze-up can occur. There are header pipes along the upper and lower edges of the array of boxes. Coolant is circulated at 20 gpm by a centrifugal pump, and the heated water flows directly into the storage system: there is no heat exchanger. Automatic tracking is provided by a single pair of photocells and a small electric motor. The twenty copper tubes are parallel to one another and never move or turn. The boxes rotate about their respective tubes as axles. The total range of tracking, from east to west, is 180 degrees. But during heavily overcast days, and at night, the boxes are automatically oriented downward for protection from rain, snow, hail, dust, and vandals. This same orientation is used if, on a sunny day, the temperature of the coolant for some reason exceeds 190°F.

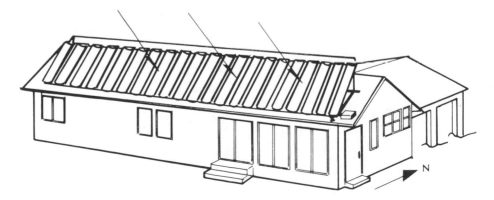

Storage   The 2000 gallons of water and inhibitor is kept in two tanks of 1000-gallon capacity each. Each tank is a vertical, cylindrical steel tank 5½ feet in diameter and 6 feet high, insulated with 6 inches of rockwool. The tanks are side by side in the utility room in the northeast portion of the basement. The rooms are heated by baseboard radiators.

Auxiliary Heat Source   Gas-fired water heater.

Domestic Hot Water   This is preheated in a 42-gallon tank within one of the above-mentioned tanks.

Cooling in Summer   A conventional evaporative air conditioner is used. Cool water is stored in one of the above-mentioned large tanks. The 24-inch eaves help exclude summer sun from the south windows.

Solar heating planner, owner, occupant: John Kusianovich. *Advisor on solar heating and architecture:*   The Architects Taos. *Consultants:*   Burns and Peters. *Designer of collector system and control system:*   W. Myers. *Manufacturer of collector and general contractor:*   Albuquerque Western Solar Industries, Incorporated. *Funding:* Private. *Cost of collector: Delivered but not installed:*   About $7 per square foot.

Sandia Heights House

Albuquerque   35°N
660 Roadrunner
Lane, in Sandia
Heights

> **Building:** 2-level, 2200 sq. ft.
> **Collector:** 640 sq. ft., concentrating type
> **Storage:** 1000 gals. water
> **% Solar-heated:** 75 (predicted)

Two detached arrays of cylindrical, parabolic collector boxes provide 75% solar-heating.

Building   This is a two-level, three-bedroom, wood-frame house with no attic. Beneath the main story there is a two-car carport and a recreation room. Window area is moderate, and all windows are double-glazed. The windows on the south side are deeply recessed to exclude summer sun. The walls are insulated with 4 inches of fiberglass and the ceiling or roof contains 10 inches. The house faces west.

Collection   The collector is mounted, in two arrays, on two 60-by-8-foot steel stands situated about 50 feet northeast of the house. The arrays face straight south and slope 45 degrees. Each array includes 20 Solcan collector boxes, made by Albuquerque Western Solar Industries, Incorporated; there are forty boxes in all. Each is 8 by 2 feet, with the long axis pointing upward to the north and downward toward the south, at 45 degrees from the horizontal. Automatic tracking is provided. The coolant is water. No antifreeze is used; the liquid is drained before freeze-up can occur; and it is drained whenever tracking is discontinued. The water is circulated through the collector at 40 gpm by a 1-hp centrifugal pump. The hot water flows directly into the storage system; there is no heat exchanger.

Storage   The 1000 gallons of water is contained in a horizontal, cylindrical, coated steel tank buried in the ground at the east end of the

house. The tank is insulated with 3 inches of urethane foam. The rooms are heated by a forced-air system.

Auxiliary Heat Source   Electric heating elements in the forced-air duct.

Domestic Hot Water   This is preheated in a 40-gallon steel tank within the above-mentioned tank. Final heating, if needed, is provided by electric heating elements in a standard 60-gallon hot water heater tank.

Cooling in Summer   Evaporative cooling is used.

Problems and Modifications   Initially, a fiberglass storage tank was used. It soon ruptured and was replaced by a coated steel tank.

Solar engineer and manufacturer of the solar heating system: Albuquerque Western Solar Industries, Incorporated. *Architect:* Burns and Peters. *Builder:* Sandia Homes, Incorporated. *Cost of house proper:* $66,500. *Cost of solar heating system:* $7000.

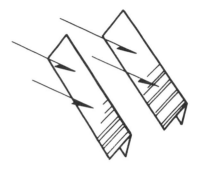

Collector arrays

Albuquerque   35°N

Building: 3-story, 3000 sq. ft.
Collector: 1000 sq. ft., water type
Storage: 3000 gals. water
% Solar-heated: 85

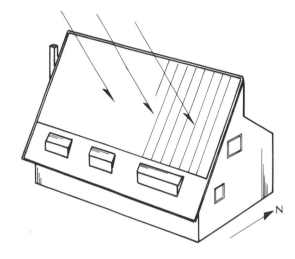

Three-story, 3000-sq.-ft. house with 1000-sq.-ft. water-type collector

Building   This is a four-bedroom wood-frame house with detached two-car garage. The walls are insulated with 4 to 6 inches of fiberglass and the roof contains 6 inches. There are six south-facing windows, each 32 by 24 inches, and smaller window areas on the other sides of the house. All of the windows are double-glazed.

Collection   The 46-by-28-foot collector is on the south roof, which slopes 55 degrees. The fifty-one collector panels are of a special design developed by the owner. Each is 100 by 15 inches and includes an aluminum Roll-Bond sheet, made by Olin Brass Company, that has eight integral passages for coolant. The black surface is nonselective. The glazing is double and consists of sheets of Filon (polyester and fiberglass, with Tedlar coating). The coolant is water to which Nalco 39P inhibitor has been added. No antifreeze is used. The system drains automatically before freeze-up can occur. The coolant is circulated at 26 gpm by a ⅓-hp centrifugal pump and then flows directly into the storage system. The panels are arranged in seventeen adjacent columns of three panels each. The tubes within the individual panel are hydraulically in parallel, the three panels of a column are in series, and the seventeen columns are in parallel.

Storage   The 3000 gallons of water is in a steel-reinforced, poured-concrete tank under the house. The tank has a waterproof fiberglass

liner. Rooms are heated by a forced-air system that makes use of the furnace blower.

Auxiliary Heat Source Gas-fired hot-air furnace. Also a fireplace equipped with a heat exchanger.

Domestic Hot Water This is preheated by the solar heating system.

Cooling in Summer There is no formal cooling system. Little cooling is needed inasmuch as the house is well insulated and the window area is small.

Garage Heating The two-car garage, situated 50 feet to the north, has its own solar heating system. The air-type collector is mounted on the sloping roof of the garage.

Problems and Modifications When the water in the concrete storage tank was heated to 175°F, the tank cracked. The liner, however, was unaffected.

Solar engineer, owner, occupant: Frank Zanner. *Architect:* Don P. Schlegel. *Builder:* Manzano Mountains Construction Company. *Funding:* Private.

Corrales   35°N
(suburb of
Albuquerque)

Building: 10 small 1-story rooms
Collector: }
Storage: } Passive
% Solar-heated: 90

Ten-zome house 90% solar-heated by four window-walls and four stacks of water-filled drums

Building This one-story, rambling-style house consists of a curved chain of ten single rooms connected by doorways so that air can circulate from one to another except when the doorway curtains are closed. Many of the rooms are not rectangular parallelepipeds but have ten or more faces; these rooms are called zomes. Many of the walls are of adobe and many of the floors are 5-inch concrete slabs. There is no basement or attic.

Passive Solar Heating System The south walls of the four south rooms are vertical windows, each about 10 by 10 feet in area, that are single-glazed with single-strength window glass. Behind each of the four window-walls are rows of 55-gallon steel oil drums, stacked four or five high, one above the other, in steel support frames. Each drum is oriented with its axis horizontal, north-south, and is filled 95 percent full with water. There are ninety drums in all, and they contain 20 tons of water. The south ends of these drums are coated with a nonselective black paint. At night, each of the four large window areas is covered by a large, flat, aluminum-faced insulating cover, hinged at the bottom, that is raised by handcrank and 3/8-inch-diameter nylon rope. During sunny days the cover rests horizontally on the ground and acts as a crude mirror to direct additional radiation to-

ward the window. Inside the rooms, curtains partially control the flow of radiant and convective heat from the drums to the rooms. The adobe walls and concrete floors contribute to thermal storage, about half of which is provided by the water-filled drums. Some of the rooms have skylights that are equipped with automatically operated, solar-powered, thermally insulating Skylids. The system as a whole provides about 90 percent solar heating if temperature fluctuations throughout the 55° to 80°F range are accepted.

Auxiliary Heat Source   Wood-burning stoves.

Domestic Hot Water   This is solar-heated by two 40-square-foot passive (thermosiphon) collectors adjacent to the southeast and southwest corners of the building.

Cooling in Summer Cool night air is allowed to circulate past the drums, cooling them. During the day the drums cool the rooms. Room air can be kept at 80°F even when the outdoor temperature reaches 100°F. The big insulating covers are kept closed a large part of the time in summer.

Problems and Modifications During seven years of use the system has operated uneventfully. Two zinc-bunged drums developed leaks at the bung, but most of the drums are steel-bunged and have not leaked. The drums sometimes "bong" when large changes in temperature occur.

Designer, owner, occupant: S. C. Baer. *Cost of drums:* $2 to $4 each if second-hand, $9 if new.

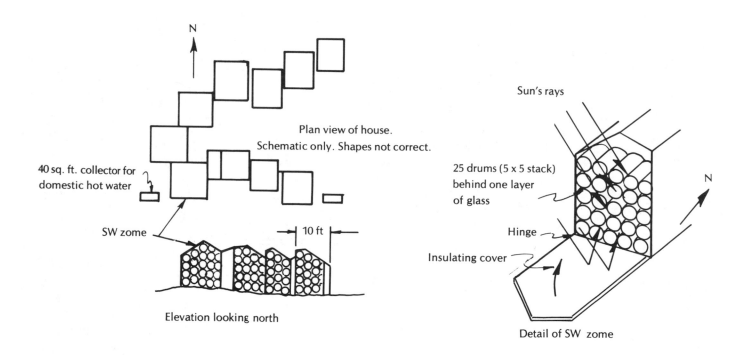

N

Plan view of house.
Schematic only. Shapes not correct.

40 sq. ft. collector for domestic hot water

SW zome

10 ft

Elevation looking north

Sun's rays

25 drums (5 x 5 stack) behind one layer of glass

N

Hinge

Insulating cover

Detail of SW zome

25,000-sq.-ft. laboratory building, 78% solar-heated and 45% solar-cooled

Las Cruces 32° N
(200 mi S of
Albuquerque)
On NMSU campus

Building The rectangular one-story building houses a laboratory for the study of crop diseases and pesticides. The plan dimensions are 177 by 128 feet. The floor area is 25,000 square feet gross, 19,000 square feet net. There is a small basement for mechanical equipment. The walls are of vermiculite-filled 12-inch concrete blocks insulated with 1 inch of urethane foam and sheetrock sheathing. Ceilings are insulated with 6-inch fiberglass batts, and one inch of rigid insulation is incorporated in the roof. The window area is small, amounting to only 5 percent of the wall area. All windows are double-glazed. The rate of exhaust of air from laboratory hoods is 5900 cfm; a heat exchanger recovers much heat from the exhaust air.

> **Building:** 1-story, 25,000 sq. ft.
> **Collector:** Two types used; see text
> **Storage:** 28,000 gals. water
> **% Solar-heated:** 78

Collection The collector panels are arranged in eight east-west rows on the horizontal roof. The panels slope 30 degrees. The 6800-square-foot collector consists of two portions: a main portion and a supplementary portion.

The main portion, with an area of 6260 square feet, consists of 336 panels made by Sunsource Division, Daylin, Inc. Each panel is 6 by 3 feet. The absorber is a sheet of 20-gauge mild steel to which galvanized steel tubes, ½ inch in inside diameter and 5 inches apart, have been mechanically clipped. The a/e ratio of the Tabor-type se-

lective black coating is 0.9/0.1. The glazing is single and consists of 5/32-inch water-white glass. The panel backing includes 4 inches of fiberglass. The coolant is water to which 40 percent of propylene glycol has been added. A flowrate of 0.04 gpm per square foot of collector area is maintained by two 3-hp centrifugal pumps. The heat is delivered to the storage system by a heat exchanger.

The supplementary portion, with an area of 540 square feet, consists of 54 Northrup, Inc., tracking collector boxes which provide a nominal 10-to-1 concentration. Each box has an aperture of 10 feet by 1 foot, which is covered by a cylindrical-type methyl methacrylate fresnel lens with 14-inch focal length. The long axis of the box slopes 30 degrees upward toward the north in winter, but it can be manually adjusted for other seasons. The absorber proper is a 2-inch-wide copper strip to which a 3/8-inch copper tube has been welded. The black coating on strip and tube is nonselective. The coolant used, and the flowrate, are the same as for the main portion of the collector. The heat collected is sent to the same storage system.

Storage The 28,000 gallons of water is in two 14,000-gallon, pressurized, horizontal, cylindrical steel tanks situated below ground near the southeast corner of the building. Each tank is insulated externally with 3½ inches of urethane foam. Heat is distributed to the rooms by two duct-and-fan-coil systems.

Auxiliary Heat Source Gas-fired boiler.

Domestic Hot Water This is fully heated by the solar heating system.

Cooling in Summer Two absorption-type coolers, employing lithium bromide, are used: a York 155-ton cooler derated to 50 tons and and Arkla 25-ton cooler derated to 16 tons. These are fed with water at 175° to 185°F supplied from either portion of the collector or from the auxiliary heat source. The percent solar cooling is 45.

Solar engineering: Bridgers and Paxton; also New Mexico State University Research and Development Institute (R. L. San Martin and others). *Architect:* W. T. Harris and Associates. *Builder:* Wooten Construction Company. *Coordination:* NMSU Mechanical Engineering Department and Physical Sciences Laboratory. *Cost:* $1,500,000. *Funding:* By State of New Mexico.

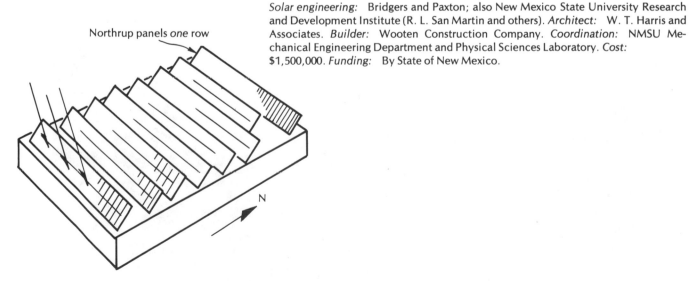

Northrup panels *one row*

N

A 59,000-sq.-ft. building, with 8000-sq.-ft. collector employing oil-type coolant, that is 95 % solar-heated and 70 % solar-cooled

Los Alamos 36° N

Building   Housing the National Security and Resources Study Center and also the main library of the Los Alamos Scientific Laboratory and a special reports library, this building has 59,000 square feet of heated area and 6600 square feet of unheated utility space. The walls and roof are well insulated, a typical R-value being 19. All windows are double-glazed. Many design features contribute to reducing heat loss in winter and limiting heat intake in summer. A heat-recovery system is used in winter. Heating and cooling requirements are reduced by use of a floating temperature-and-humidity setpoint. The annual heat load is 2.5×10^9 Btu.

Collection   The liquid-type collector occupies a 100-by-80-foot area of a south roof that slopes 35 degrees. There are four-hundred panels of LASL design which, besides collecting solar energy, serve as the roof. Each panel is 10 by 2 feet. The heart of a panel is a pair of ungalvanized steel sheets that have been placed face to face and welded together in a certain pattern of areas and then pressure-expanded to provide many parallel passages for flow of coolant liquid. The passages are so thin that the average thickness of coolant is only 0.035

Building:	3-story, 59,000 sq. ft.
Collector:	8000 sq. ft., liquid type
Storage:	15,000 gals water
% Solar-heated:	95 (predicted)

inch and accordingly the thermal capacity (and warm-up time) is small. The edges of the upper sheet are bent upward, and the edges of the lower sheet downward, to form an integral frame (see drawing) that supports, above the upper sheet, the metal strips that hold the single layer of 1/8-inch glass, and below the lower sheet, a 3½-inch layer of fiberglass and/or urethane insulation. The frames of adjacent panels are clipped together by inverted-U-section clipping strips to form a strong, watertight array. The black chrome coating is selective: the a/e ratio is 0.95/0.1. The coolant is the light paraffinic oil Shell Thermia 33, and the flowrate is 420 gpm. Heat is transferred to the storage system via a heat exchanger. The collector array is supported by horizontal steel I-beams 10 feet apart.

Storage   The 15,000 gallons of water is stored in two tanks: a 10,000-gallon tank at low pressure and a 5000-gallon tank at high pressure. The rooms are heated by forced hot air; fan-coil systems are used. There are two main heating zones and these are separately controlled.

Auxiliary Heat Source   The central steam-supply system of the Los Alamos Scientific Laboratory.

Cooling in Summer   In summer the 10,000-gallon tank is used for cold storage. Two solar-powered cooling systems are used: a York absorption-type chiller and a Rankine-cycle chiller made by Barber-Nichols Engineering Co. Either can be used to cool the cold-water tank, and either can be used to cool the rooms directly. Percent of cooling power supplied by solar system: 70% (predicted).

Solar engineering: Los Alamos Scientific Laboratory of the University of California (J. D. Balcomb, J. C. Hedstrom, S. W. Moore, et al.). *Solar consultant:* Jerome Weingart. *Architect:* Charles Luckman Associates (esp. Sam Burnett). *Engineer:* Ayres and Hayakawa. *Contractor:* John C. Cornell. *Owner and source of funds:* Energy Research and Development Administration. *Cost of building and solar heating and cooling system:* $4,6000,000 (est.). *Cost of solar heating and cooling systems alone:* $440,000 (est.)

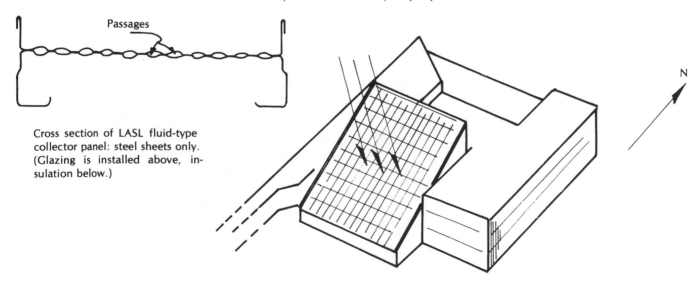

Cross section of LASL fluid-type collector panel: steel sheets only. (Glazing is installed above, insulation below.)

Passages

N

Nambe Indian Pueblo Community Building

Air-type collector in long slender penthouse is assisted by a large-area aluminum reflector

Nambe   36½° N
(alt. 6200 ft.; a 5300-degree-day-site)
(20 mi N of Santa Fe)

Building This one-story, 3080-square-foot addition to an existing building embodies traditional Southwestern architecture and includes tribal offices, meeting areas, a classroom, and a library. A long slender penthouse extends along the northeast part of the horizontal roof. There is no attic, basement, or garage. The 14-inch-thick adobe walls are insulated on the outside with 2 inches of polystyrene foam protected by a thin layer of stucco-on-mesh. The partition walls are of 8-inch-thick pumice blocks. The floor is a 4-inch concrete slab. The window area is small, and the windows are recessed and double-glazed. The building faces 45 degrees west of south. The thermal load, including losses from air leakage, is 18,600 Btu per Fahrenheit degree-day (predicted).

Collection The collector occupies the southwest portion of the penthouse. The vertical window, 66 by 6½ feet, is single-glazed with 3/16-inch tempered glass. The heart of the collector is a set of four 66-foot-long horizontal ducts, each with a cross section of 18 by 3½ inches. The ducts are mounted on edge, one above the other, so as to occupy the entire area of the window. The ducts are of .018-inch galvanized steel and have a nonselective black coating. Air flows horizontally northwest within the top and bottom ducts, returning through the two intervening ducts. Both headers (risers) are at the southeast end of the penthouse. The insulation behind the array of ducts includes 2 inches of rigid fiberglass and a 6-inch fiberglass batt.
Reflector: A horizontal roof area 72 by 16 feet, adjacent to the penthouse, is covered with Coilzak, an aluminum sheet that has high reflectance (about 80 percent) and has been anodized to reduce

> **Building:** 1-story, 3080 sq. ft.
> **Collector:** 440 sq. ft., air type
> **Storage:** 17-ton bin-of-stones
> **% Solar-heated:** 65 (predicted)

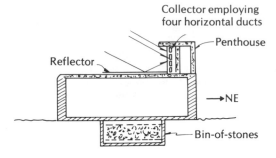

Collector employing
four horizontal ducts

Reflector

Penthouse

→NE

Bin-of-stones

Cross section of building
looking NW

corrosion. The sheet is backed by a 1½-inch layer of Styrofoam insulation. In winter this reflector considerably increases the amount of radiation incident on the collector. Hot air from the collector is circulated at 880 cfm, by a ¾-hp blower situated in the penthouse, to the storage system.

Storage   The 17 tons of 2-to-3-inch-diameter stones is in a rectangular, 12-by-12-foot, 4-foot-high bin situated beneath the floor of the building. Distribution of air at the bottom of the bin is by eleven perforated 6-inch-diameter corrugated metal pipes such as are commonly used for highway drainage. Hot air from the collector flows downward through the bin, and cold air from the rooms is circulated (by the furnace blower) upward through the bin. The massive walls and floor of the building contribute greatly to thermal storage.

Auxiliary Heat Source   Gas furnace.

Domestic Hot Water   This is not solar-heated. Little hot water is needed.

Cooling in Summer   At night, outdoor air is cooled by evaporation of sprayed water and is then driven through the bin-of-stones, cooling it. During hot days, room air is circulated through the bin and is cooled by it. Little cooling is needed inasmuch as the building is well insulated, windows are small and are recessed, and the walls and floor have large thermal capacity.

Solar engineering supervision: Los Alamos Scientific Laboratory of the University of California (especially Solar Energy Group Q-24, J. D. Balcomb et al.). *Architect:*   A. L. McNown. *Construction superintendent:*   F. Tixier. *Mechanical construction:* G&G heating and Sheet Metal Company. *Funding:* $150,000 from the Economic Development Administration of the Department of Commerce, for the building, and $33,000 from Energy Research and Development Administration for the solar-heating system; also $25,000 from Energy Research and Development Administration for instrumentation and performance monitoring.

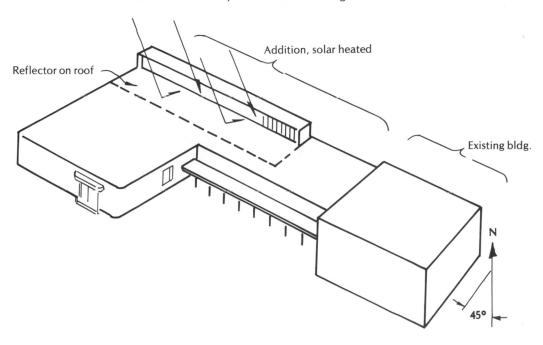

Reflector on roof

Addition, solar heated

Existing bldg.

N

45°

Benedictine Monastery Office and Warehouse

This 7700 sq.-ft. office-and-warehouse building is 95% solar-heated by an $8500 passive system employing three long rows of windows and a long row of water-filled drums.

Pecos   35½°N
(alt. 7700 ft.)
(35 mi SE of Santa Fe)

Building: 1-story, 7700 sq. ft.	
Collector: } Passive	
Storage: }	
% Solar-heated: 95	

Building The 140-by-55-foot building has a parital (3000-square-foot) basement and no attic. The floor is a 4-inch concrete slab. The walls, of 8-inch concrete blocks, are insulated on the inside with 2 inches of beadboard to R-12. The roof is insulated with 6 inches of fiberglass to R-20. The window areas on east, north, and west are small, amounting in all to 10 square feet, and are single-glazed. The south windows are described below. Offices occupy the south portion of the building, and the north portion, served by clerestory windows, is a warehouse. The building faces exactly south.

Passive Solar-Heating System Radiation is received via three areas of windows: lower area, intermediate area, and upper area; all are vertical and are double-glazed with Thermopane. The total area is 1345 square feet.

Lower area. This is 120 by 3½ feet, with a total window area of 426 square feet. It extends along the lower portion of the south side of the building. External, manually operated shutters of .050-inch alumi-

num are provided; they are kept closed in summer to exclude radiation and reduce conductive heat inflow; they can be closed on winter nights to reduce heat loss. On winter days, solar radiation passing through these windows strikes black, water-filled steel drums and is absorbed by them. There are 140 drums, each of 55-gallon capacity. Two additives, intended to prevent corrosion, are included in the water: (1) Universal Oil Products F-45 inhibitor, 8 ounces per drum, and (2) an equal amount of 30-weight motor oil. The drums are horizontal and are arranged in two rows, one above the other. The lower row rests in a 4-foot-wide, 1½-foot-deep depression; the tops of the upper-row drums are 3 feet above floor level. The drums are in openable housings insulated with 4 to 6 inches of fiberglass.

Intermediate area. This is 140 by 3½ feet, with a total window area of 406 square feet. These windows are situated directly above the previously described windows. Besides providing occupants with natural illumination and a view of outdoors, the intermediate window area admits solar radiation deep into the room where is strikes and warms the concrete floor-slab and the massive east-west partition wall.

Upper area. This is 140 by 4 feet, with a total window area of 513 square feet. Radiation entering here lights and warms the north part of the building. A light-colored, sloping roof reflects additional radiation toward this window area.

Auxiliary Heat Source Nine manually controlled 1.4 kw radiant electric heating panels.

Domestic Hot Water This is heated by a separate 40-square-foot collector.

Cooling in Summer The lower area of windows is covered by aluminum shutters. The clerestory windows are opened to allow venting of hot air. The large thermal capacity of water-filled drums, floor, partition wall, etc., reduces the temperature rise.

Solar engineering and architectural design: Zomeworks Corporation (S. C. Baer, et al.). *Builder:* M. Hanson. *Owner and occupant:* Dove Publications of Benedictine Monastery. *Cost of building and solar heating system:* $130,000, or $13 per square foot. *Cost of solar heating system:* $8500.

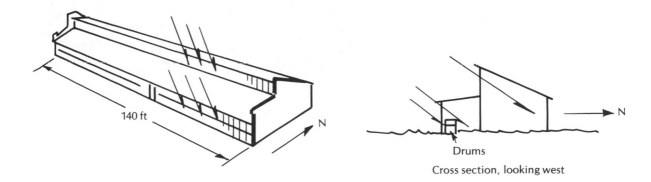

140 ft

N

Drums

Cross section, looking west

Passive solar heating system employing large window area and massive walls and floor

Santa Fe   36°N
(alt. 7500 ft.)

Building   This one-bedroom house, 65 by 18 feet, has no attic, basement, or garage. It faces 7 degrees east of south. The south wall includes ten 6-by-4-foot windows, the west wall has one such window, and the east wall has two 3-by-3-foot windows. The north-wall window area is negligible. All windows are double-glazed (Thermopane). The windows are covered at night with magnetically affixed 1-inch-thick plates of polystyrene foam. The south-wall areas between windows are of 2-foot-thick adobe, and the north wall also is of 2-foot-thick adobe. All walls are insulated on the outside with 2 inches of urethane foam protected by a layer of stucco cement. The floor consists of a single layer of 2¼-inch-thick indoor-type bricks laid on a bed of sand. There are no rugs on the floor.

Passive Solar Heating System   The large windows on south, east, and west sides, with a combined area of 280 square feet, admit much solar radiation, which strikes walls and floor. Light-colored sand has been spread on the ground adjacent to the south side of the building to reflect radiation toward the windows. Heat storage is provided by the massive walls and floor.

Auxiliary Heat Source   Fireplaces in living room and bedroom. Also, electric heating mats embedded in the sand at a depth of 1 ¾ inches.

Domestic Hot Water   This is not solar-heated.

Cooling in Summer   None needed, none provided.

Solar engineer: Jay Shelton. *Architect:* William Lumpkins. *Builder:* Aztec Construction Company. *Owner and occupant:* A. V. Dasburg. *Cost of house and solar-heating system:* $45,000.

Building: 1-story, 1200 sq. ft.
Collector: Passive
Storage:
% Solar-heated: 55

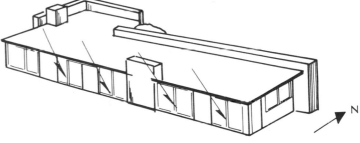

Santa Fe   36°N
1002½ Canyon Road

Building: 1-story, 500 sq. ft.
Collector: 230 sq. ft., air type
Storage: Complex; see text
% Solar-heated: 65

Small house with air-type solar heating system is now in its twentieth year

Building   An old one-story adobe building, built about 1935, was extensively remodeled to provide partial solar heating. There is no basement or attic. The building is 32 feet long and 16 feet wide. The walls are of adobe and are 12 to 14 inches thick. The roof is horizontal and is insulated. The windows are small and are double-glazed. The long axis of the house is north-south, which complicated the task of designing a solar heating system.

Collection   The 230-square-foot air-type collector is in two parts, the north part being 50 percent larger than the south part. Both parts slope 45 degrees and both are single-glazed with double-strength glass. One inch below the glass there is a galvanized steel sheet that has a nonselective black coating. The underside of the sheet is finned, and the airflow is in the space containing the fins. The 1-inch space between steel sheet and glazing is dead airspace. Airflow is maintained by 80-watt blowers and 5-inch-diameter ducts leading to the storage system. There are two blowers and two ducts for the north collector and one of each for the south collector. The blowers run whenever a differential thermostat shows that the collector temperature exceeds the storage temperature by at least 12 Fahrenheit degrees. The underside of each collector is insulated.

Storage   Many components and materials contribute to the storage of heat. In the north portion of the building much heat is stored in a 12-inch-thick bed of sand below the brick floor. Several parallel, horizontal, 3-inch-diameter pipes (downspout pipes), 16 inches apart on centers, are buried in the sand bed. These pipes receive hot air from the collector and discharge the air to the south part of the room. The

air to be returned to the collector is air from close above the north end of the floor. The edges of the sand bed are insulated with 1½ inches of Celotex insulation. In the south portion of the building there is a 4-foot-deep pit of fist-size stones, with 8-inch pumice blocks along the sides for insulation. Sheet metal resting on the stones supports the sand and the brick floor. Room air is circulated through the stone-filled pit by an 80-watt blower. The thick adobe walls of the building contribute to thermal storage.

Auxiliary Heat Source   A high-efficiency fireplace and a small (12,000 Btu/hr) gas heater.

Domestic Hot Water   Preheated in helical pipe coil inserted in the duct from the south collector.

Cooling in Summer   None provided. Little is needed because of the effective exclusion of radiation and the large thermal capacity of walls and floors.

Problems and Modifications   Few difficulties have arisen in nineteen years of use. Two glazing sheets of the collector broke and were replaced. The designer believes the performance might have been better if the collectors were vertical (to shed snow and to permit use of horizontal reflectors) and if, in the northern part of the building, a bin-of-stones had been used instead of ducts-in-sand.

Solar engineer and owner: Peter van Dresser.

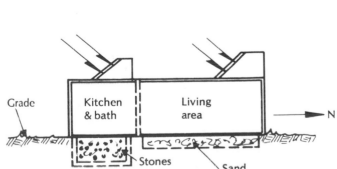

Vertical cross-section, looking west

179

Santa Fe   36°N
(alt. 7000 ft.; a
5900-degree-day site)

Building: 2-story, 1900 sq. ft.
Collector: ⎰ Combination active-and-
Storage: ⎱ passive system
% Solar-heated: 80

A large two-story greenhouse in south-central part of house serves as a walk-in collector. Storage is partly passive, partly active

Building   The living quarters constitute an L-shaped structure, with a two-story greenhouse occupying the south-central space. There is no basement or attic. The exterior adobe walls are 10 to 14 inches thick, and the interior adobe walls that face the greenhouse and surround the stairwell are 14 inches thick. The roof includes 3 inches of urethane foam. The windows are double-glazed. The building is set 3 feet below grade and faces exactly south.

Combination Active-and-Passive Solar Heating System   The greenhouse serves as a walk-in collector. Its window area is 400 square feet. About a third of this is vertical and the rest slopes 60 degrees. Double glazing is used throughout. The solar radiation that enters is absorbed by the massive floor and adobe walls. Hot air that accumulates in the uppermost part of the greenhouse is exhausted by blow-

ers, through vertical ducts, to two bins-of-stones buried beneath the floor at either end of the greenhouse. Together, these bins contain 900 cubic feet (about 50 tons) of 3½-inch-diameter stones. The bins are insulated. When the rooms need heat, room air is circulated through the bins.

Auxiliary Heat Source   Three-stage 10-kw electric heater in duct from bin.

Domestic Hot Water   This is preheated by a special 34-square-foot solar collector mounted on the horizontal roof. Heat is delivered to a 40-gallon tank by a heat exchanger.

Cooling in Summer   The stones in the bins can be cooled by forced circulation of cool night air. The large mass of the building prevents any large rise in room temperature.

Architect: Sun Mountain Design. *Engineer:* Communico. *Builder:* Wayne Nichols, Communico.

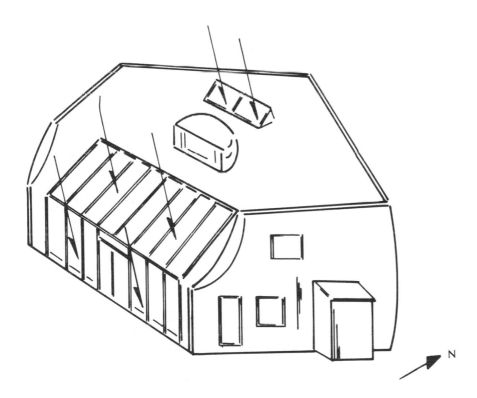

Fitzgerald House [Suncave]

Santa Fe   36°N
(alt. 6900 ft.; a 5900-
degree-day site)

Building:	1-story, 1500 sq. ft.
Collector:	} Passive
Storage:	
% Solar-heated:	95

Adobe-and-concrete house built into side of hill and 95% passively solar-heated via 440 sq. ft. of vertical south windows

Building   This is a one-story, two-level, two-bedroom adobe-and-concrete house built into a south-facing hillside. There is no attic, basement, or garage. The roof is covered with sod. The adobe walls, 10 inches thick, are insulated on the exterior with 2 inches of urethane foam. The brick floors rest on a 16-inch-thick bed of sand which rests in turn on a 1-inch layer of Styrofoam. The windows are double-glazed.

Passive Solar Heating System   Radiation enters via 440 square feet of vertical, double-glazed windows. Much heat is stored in the massive walls and floors. No thermal shutters or shades are used at night.

Auxiliary Heat Source   Two fireplaces. There is an electric heater in the bathroom.

Domestic Hot Water   This is not heated by the solar heating system.

Cooling in Summer   The 1½-foot eaves exclude much radiation. The well-insulated and massive walls and the sod-covered roof transmit little heat. The massive walls and floors reduce temperature rise. Cross-ventilation is provided.

Solar engineer: Mark Chalom and David Wright. *Architect:*   David Wright. *Builder:* K. E. Terry. *Owner, occupant:*   D. Fitzgerald. *Funding:*   Private.

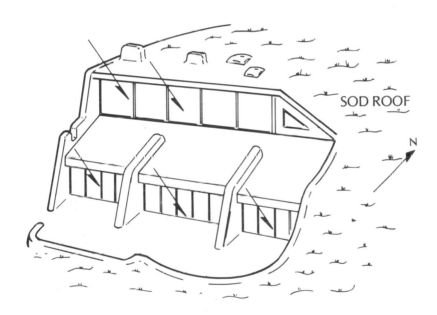

SOD ROOF

N

Sunscoop [Kimball] House

Santa Fe   36°N

Building: 1½-story, 1100 sq. ft.
Collector: }
Storage: }   Passive
% Solar-heated: 90

A 32-by-12 1/2-foot vertical window-wall provides 90 % of the heat needed by this 1100-sq.-ft. adobe house

Building   The 1100 square feet of living space includes a loft used as sleeping quarters. There is no basement or attic. The adobe walls, 10 to 14 inches thick, are insulated on the exterior with 2 inches of Styrofoam protected by a 1-inch layer of adobe-and-cement. The effective thickness of first-story floor is 24 inches.

Passive Solar Heating System   Solar radiation is received via a vertical, double-glazed, 400 square-foot south window-wall 32 feet long by 12½ feet high. The transmitted radiation strikes the massive floor and walls, warming them. Canvas-and-Styrofoam shutters, closed at night, reduce energy loss through the window wall.

Auxiliary Heat Source   A wood-burning stove. About half a cord of wood is burned per year.

Domestic Hot Water   This is heated by a special water-type collector adjacent to the southeast corner of the house. The collector is 8 by 4 feet and slopes 45 degrees. The coolant (water and ethylene glycol) flows by gravity convection to an 80-gallon, glass-lined storage tank situated within the house. The collector was made by Zomeworks Corp.

Cooling in Summer   Eaves that project 4 feet exclude much radiation. Hot air within house escapes via vents.

Problems and Modifications   Initially, overheating occurred at some times of year. Such overheating was reduced by installation of additional shades.

Solar engineer, architect, and original owner: David Wright. *Subsequent owner:* Clark Kimball. *Funding:*   Private

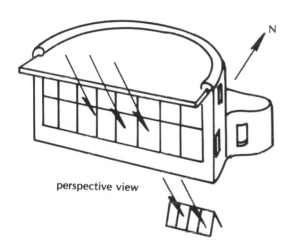

perspective view

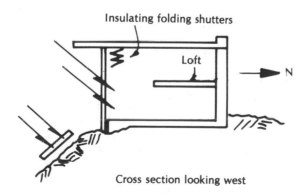

Insulating folding shutters

Loft

N

Cross section looking west

Terry House

Santa Fe   36°N
(alt. 7000 ft.; a 5900-
degree-day site)

Building: 1-story, 850 sq. ft.
Collector: }
Storage: }   Passive
% Solar-heated: 90

House that is stepped to conform to slope of hill and is passively so-lar-heated by four skylights, one per step

Building   This one-story, one-bedroom, adobe-brick house has three levels, to conform to the slope of the ground. At the north end there is a small loft. The entire structure is insulated on the outside with 2 inches of rigid urethane foam. There are earth berms on the east, north, and west sides. The adobe walls are 14 inches thick. The house faces exactly south.

Passive Solar Heating System   Radiation enters via four south-slop-ing skylights, each 16 by 6 feet. The total area is 380 square feet. Each skylight consists of four side-by-side, 76-by-46-inch windows double-glazed with tempered glass and insulated at night by internal shut-ters. Thermal storage is provided by the building itself and by 1100 gallons of water. The water is in twenty-two 55-gallon steel drums, each having a thin coating of adobe plaster. The drums are mounted on end and arranged in rows to form room dividers and are situated so as to receive, in winter, much direct radiation via the skylights. The thermal capacity of the set of drums is 9000 Btu per Fahrenheit degree. Much thermal capacity is provided also by the 65 tons of adobe walls and the 48 tons of adobe floors. Heat is imparted to the rooms by radiation and convection from the drums, walls, and floors.

Auxiliary Heat Source   Two wood-burning stoves.

Domestic Hot Water   This is not heated by the solar heating system.

Cooling in Summer   External shutters, or louvers, are installed over the skylight areas. Natural ventilation is used. The large thermal capacity of the building and the drums tends to prevent the rooms from becoming very hot.

Solar engineer and architect: David Wright. *Owner:*   K. E. Terry. *Funding:*   Private.

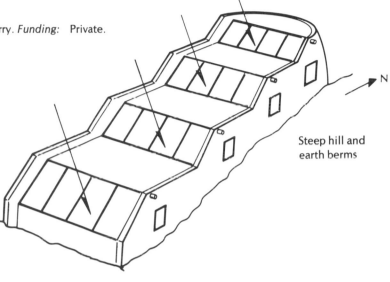

Steep hill and earth berms

Egri House

Taos
In Las Colonias area,
just N of El Prado

Building:	1-story, 925 sq. ft.
Collector:	} Passive
Storage:	}
% Solar-heated:	About 80

Small adobe house heated by low-cost, passive, add-on collection-and-storage room with removable plastic double-glazing.

Building   This [existing,] one-story, two-bedroom adobe house was built in about 1909. The house has no attic and no garage. One bedroom is used as a studio. Only two rooms are regularly heated: the kitchen-dining-living room and the main bedroom. Their combined area is 350 square feet. The walls are of thick adobe and have no added insulation. Window areas are small. Adobe earth rests on the roof, insulating it. The house faces 18 degrees west of south. The radiation-receiving enclosure, described below, has insulated concrete footings.

Passive Solar-Heating System   Solar radiation is collected by a special radiation-receiving enclosure, added in 1975, abutting the center of the southwest face of the house. The enclosure is 16 feet long, 8 feet deep, and 7 feet high. Its top and its southwest face are transparent, being double-glazed with .006-inch Monsanto plastic film #602 supported by curved steel pipes 18 inches apart on centers and held taut by weights at lower edges. The two films are 2 inches apart. The transparent area is 160 square feet. Radiation entering the enclosure strikes the earth floor and the vertical, brown-painted adobe wall, heating these and also heating the adjacent air. Along the base of this wall (immediately south of it) there are two rows of water-filled 20-gallon tanks (twenty tanks in all) which contribute to

the thermal storage of the system. Each tank is a sealed vertical cylinder, made of cardboard and plastic, that has been painted brown. Hot air is supplied to each of the two heated rooms of the house by gravity-convective flow from the radiation-receiving enclosure; the flow is via an upper (shoulder-height) 5-by-3-foot opening in the adobe wall and a lower 14-by-12-inch opening near the floor. There are four openings in all: two for each of the rooms in question. The opening can be closed if the rooms become too hot. Plants are sometimes grown in the radiation-receiving enclosure; accordingly, it is sometimes called a greenhouse. The end walls of the enclosure, and also the bottom part of the south face, are of wood and are insulated.

Auxiliary Heat Source   Propane gas heater.

Cooling in Summer   None needed, none provided. Note: At the start of summer, the plastic glazing of the south enclosure is removed and stored until fall.

Problems and Modifications   The initially installed devices for securing the plastic glazing to the wall of the house proper were inadequate; more positive attachments, employing screws, have been installed. After two years of use the plastic glazing became brittle and failed. It was replaced by unused film from the original roll. During warm, sunny periods in the spring and fall, the temperature in the enclosure sometimes rises as high as 120°F; this is acceptable. A fan could be used to vent the hot air and reduce the temperature rise; such a fan *was* used in 1976 but the decrease in temperature did not appear to justify the added complication, added cost, or added noise.

Basis of solar design: Stephen R. Kenin. *Design modifications:*   Fred Hopman and A. S. Hobbs. *Owner and occupant:*   K. Egri and T. Egri. *Cost of materials for solar heating system:*   About $1500.

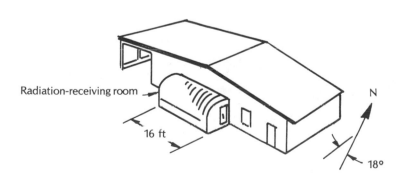

Radiation-receiving room

16 ft

N

18°

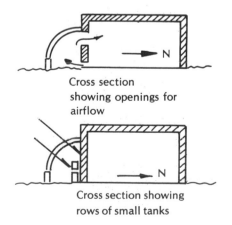

Cross section showing openings for airflow

Cross section showing rows of small tanks

Newton House

White Rock   36°N
(alt. 6700 ft.; a 6000-
degree-day site)
(near Los Alamos and
Santa Fe)

Building: 1-story, 3350 sq. ft.
Collector: ⎱
Storage: ⎰   Passive
% Solar-heated: 85 (predicted)

Large adobe house 85% passively solar-heated by three kinds of collection areas

Building This is a four-bedroom, adobe-brick building with a central greenhouse and a built-in two-car garage. The floors are of bricks resting on sand. The massive walls include 3 to 4 inches of external insulation.

Passive Solar-Heating System There are three large glazed areas. (1) The greenhouse, in the center of the building, receives solar radiation via a 600-square-foot roof that is double-glazed with plastic and slopes 15 degrees. Beneath the glazing is an array of Zomeworks Corporation Skylids which open and close automatically according to the amount of radiation incident. (2) Clerestory areas are served by 620 square feet of windows that slope 60 degrees and are double-glazed with glass. At night these windows are insulated by manually operated shutters. (3) The vertical, double-glazed south windows have an area of 100 square feet. The nominal value of the thermal capacity of the massive walls and floors is 180,000 Btu/°F.

Auxiliary Heat Source Baseboard electric heaters and fireplaces.

Domestic Hot Water Solar-heated by a 70-square-foot thermosiphon collector built by Zomeworks Corporation.

Cooling in Summer There is much shading and cross ventilation. Also there are vents above the greenhouse; venting is forced by two automatic 24-inch Nutone fans.

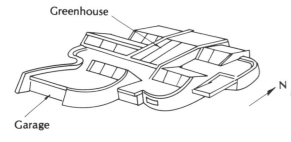

Greenhouse

Garage

N

Solar engineers: K. W. Haggard, Mark Chalom, and David Wright. *Architect:* David Wright. *Owner:* Carl Newton. *Performance monitoring:* Los Alamos Scientific Laboratory.

8000-sq.-ft. conference center employing 2300-sq.-ft. water-type collector

Albany   42° N

> **Building:** 2-story, 8000 sq. ft.
> **Collector:** 2300 sq. ft., water type
> **Storage:** 8000 gals. water
> **% Solar-heated:** 50 (see text)

Building   The first story of this wood-frame, 96-by-40-foot building includes office space, library, reception area, refreshments room, mechanical room, and storage room. The main entrance is on the south side, and the north side is recessed into the side of a hill. The second story includes a large meeting room, four conference rooms, the clerestory of the library, a kitchen, and storage room. Vertical shafts have been provided for a passenger elevator and a freight elevator, but no elevators have yet been installed. At the northeast corner of the second story there is a terrace. There is no basement and only very small attic space. The building faces 17½ degrees west of south.

Collection   The 2300-square-foot collector, 96 by 24 feet, occupies the entire south roof, which slopes 45 degrees, and is an integral part of the roof. Assembled on-site by Revere Copper and Brass Company, the collector employs a black copper sheet to which rectangular-cross-section copper tubes are affixed by means of clips and thermally conducting cement. The black coating is nonselective. The collector is double-glazed with tempered glass. The collector backing includes plywood and 3 inches of urethane foam. The coolant is water to which inhibitor has been added. No antifreeze is used; the system is drained before freeze-up can occur. The coolant is circulated at 40 gpm by a ¾-hp centrifugal pump.

Storage   Heat is stored in 8000 gallons of water contained in a steel tank buried in the ground beneath the terrace at the northeast corner of the building. The tank is cylindrical and horizontal and is insulated with 4 inches of urethane foam. Associated with the storage system is a Carrier 20-ton water-to-water heat-pump which extracts heat from the storage system and delivers the heat to the distribution system.

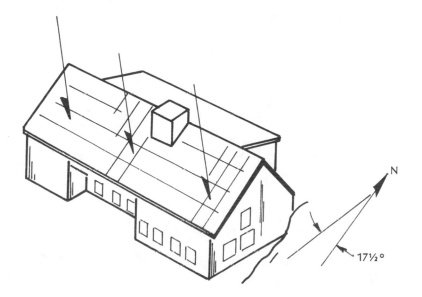

N

17½°

This system, employing three fan-coil units, circulates hot air to the rooms by way of a network of ducts. The building has eight heating zones, each with its own thermostat.

Percent Solar-Heated   The solar-heating system itself provides about 50 percent of the winter's heat need, and the combination of the solar-heating system and the heat-pump associated therewith provides 70 percent.

Auxiliary Heat Source   80 kw electric heater. A second heat-pump, intended to extract heat from outdoor air, is available also, but has been used mainly for experimental purposes.

Domestic Hot Water   This is preheated by a heat exchanger associated with the pipe carrying hot water from the collector to the storage system. Final heating is electric.

Cooling in Summer   Cooling is provided by the heat-pump that, in winter, is associated with the solar-heating system. In summer it is

operated mainly at off-peak times, and the resulting cold water is stored in a second 8000-gallon tank.

Problems and Modifications   During the first summer of operation, the collector became hotter than expected, reaching a temperature of about 325°F. The resulting thermal expansion was so great that one of the tubes connecting the collector to a header pipe became overstressed and failed. The decision was made to replace all sixteen such tubes with industrial tubes of woven bronze that have much greater compliance. Access to the collector was difficult and the changes were not completed for several months.

Solar consultant: Burt, Hill and Associates. *Technical Assistance:* Atmospheric Sciences Research Center of the State University of New York at Albany (esp. James Healey, Principal Investigator in 1977) and Niagara Mohawk Power Corporation. *Architect:* Richard Jacques Associates. *General contractor:* Reinherr and Schmidt. *Owner and user:* Alumni Association of the State University of New York at Albany. *General funding:* $230,000 by the Alumni Association of the State University of New York at Albany, $25,000 by the Atmospheric Sciences Research Center, and $65,000 by Niagara Mohawk Power Corporation. *Grant covering performance monitoring:* New York State Energy Research and Development Authority. *Cost of building complete with solar heating system:* $320,000. *Cost of solar heating system alone, including solar collector, storage system, and portion of the cost of the associated heat-pump:* $53,000.

East Hampton 41° N
(80 mi E of
New York City)

Building: 2-story, 2100 sq. ft.
Collector: 520 sq. ft., air type
Storage: 80-ton bin-of-stones
% Solar-heated: Well above 55

The 520-sq.-ft., air-type collector is in two portions, with large second-story deck between.

Building This building, with plan dimensions 44 by 36 feet, has a loft, a sundeck, and a full basement. There is no garage. The living room, kitchen, and large (36-by-12-foot) sundeck are on the second story. The bedrooms are on the first story. There is a skylight area above the south entryway. The main entrance is at the east end. In shape, the building has some resemblance to a saltbox and to a sun temple. The window areas on the east, north, and west are small. All windows are double-glazed. The walls are insulated with 3½ inches of fiberglass batts and a 1-inch layer of externally applied urethane foam protected by cedar siding. Ceilings are insulated with 6 inches of fiberglass batts and 1 inch of foam. The building faces exactly south.

Collection The air-type collector is on the sloping south roof and sloping south wall, the slope being 57 degrees in each case. There are twenty-six panels of Sunworks 1976 type. Each is 7 by 3 feet and is single-glazed with 3/16-inch tempered glass. The copper absorber sheet has a selective black coating. Between this sheet and the fiber-

glass backing there is a 1½-inch airspace in which air flows in a downward direction, driven at 2000 cfm by a ½-hp blower. On the roof there are twelve panels in a single row. The lower panels are arranged in two rows.

Storage   The 80 tons of 3-to-5-inch-diameter stones are in a steel-reinforced bin of poured concrete in the southeast portion of the basement. The bin is 18 by 14 feet by 6 feet high, and is insulated. Beneath the stones there is an 18-by-14-foot plenum. When the rooms need heat, a ¾-hp blower circulates room air through the bin.

Auxiliary Heat Source   Oil furnace with a fan-coil system.

Domestic Hot Water   This is preheated by a separate set of four panels at west end of south wall. Final heating is electric.

Cooling in Summer   None.

Problems and Modifications   In October of 1977 the rate of airflow in the collector was increased in accordance with a suggestion made by the manufacturer of the collector panels. Also the storage system was improved: the bin-of-stones was insulated and a plenum was constructed at the bottom of the bin. As a result of these improvements, the percent-solar-heated figure, which was 55 in 1976-77, is expected to be considerably higher in subsequent years.

Solar engineer, designer, builder: M. C. Matthews. *Technical advisor:*   Astral Solar Corporation (Thomas Greyson et al.). *Owner:*   Stephen Sigler.

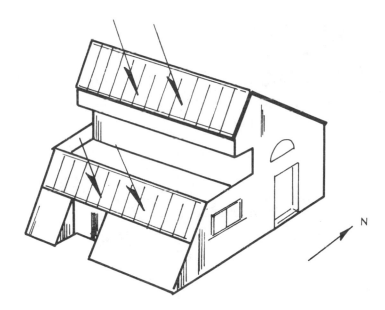

LaGrangeville   42°N
(70 mi N of New
York City)

Building: 2½-story, 4000 sq. ft.
Collector: 1200 sq. ft., water type
Storage: 4000 gals. water
% **Solar-heated:** 75

4000-sq.-ft. 75 % solar-heated by 1200-sq.-ft. water-type collector

Building: This four-bedroom, wood-frame house has a finished basement, finished attic, and, at the west end, an attached two-car garage. The walls are insulated with 5½ inches of fiberglass to R-19 and the roof is insulated with 7 inches of fiberglass to R-22. The windows are double-glazed with 1/8-inch tempered glass. Many energy-saving devices are used. The house faces exactly south.

Collection   The collector is on the central and west roofs, which slope 60 degrees. The Revere Copper and Brass Company panels form an integral part of the roof. The absorber is a copper sheet which has a selective black coating. Rectangular-cross-section tubes of type 122 copper are attached to this sheet by means of clips and thermally conducting cement. The glazing is double and consists of tempered glass. The coolant is water, with no antifreeze. The system is automatically drained before freeze-up can occur. The water is circulated at 40 gpm by a ¾-hp centrifugal pump.

Storage   The 4000 gallons of water is in a horizontal, cylindrical tank buried beneath the floor of the garage. The fiberglass tank, specially

designed by Owens-Corning, has a diameter of 6½ feet and a length of 21½ feet. The tank is insulated only by backfill of pea-size (3/8-inch) gravel, with a minimum depth of 18 inches. Heat is distributed to the rooms by a three-zone forced-air system served by three air-handlers.

Auxiliary Heat Source   Two General Electric Company air-to-air heat-pumps, of 2-ton and 3½-ton capacity, extract heat from outdoor air (if it is warmer than 15°F) and deliver the heat to a main duct. The larger heat-pump serves the first story and the smaller one serves the upper stories. Two stages of backup electric heaters come into operation automatically and sequentially when the outdoor temperature falls below 15°F and 5°F. There is a Heatilator fireplace in the living room and a standard prefabricated fireplace in the family room.

Domestic Hot Water   This is preheated in an 80-gallon tank that is immersed in the above-mentioned 4000-gallon tank.

Cooling in Summer   The above-mentioned heat-pumps are operated so as to provide cooling.

Problems and Modifications   Initially, one return line of the collector was pitched incorrectly and did not drain. The error was discovered when freeze-up occurred in this line, and the pitch was then corrected.

Architect: Harry Wenning. *Builder:* Solar Structures. *Designer of monitoring system:* General Electric Company. *Operation of monitoring system:* Central Hudson Gas and Electric Corporation. *Grant for solar-heating system:* $15,000 from HUD. *Cost of solar-heating system:* Approximately $23,000. *Selling price of house and 2 acres of land:* About $175,000.

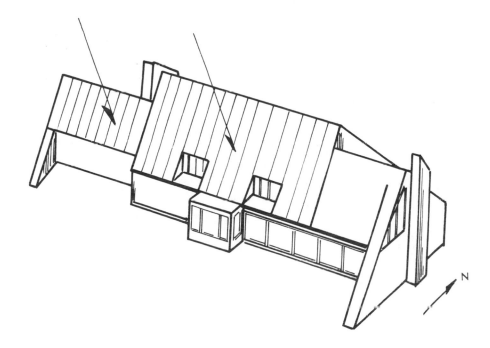

Millbrook   42°N
(a 5800-degree-day
site)
(70 mi N of New York
City)

Building: 2-story, 27,400 sq. ft.
Collector: 5650 sq. ft., water type
Storage: 15,000 gals. water
% Solar-heated: 85 (predicted)

27,400-sq.-ft. building, partly underground, 85% solar-heated by 5650 sq. ft. of water-type collectors of two different types

Building   This two-story building, of concrete and wood, is 150 by 110 feet. It is situated on a heavily wooded 1800-acre tract and faces 4 degrees west of south. It is used for research, education, and administration. The upper (main) story includes laboratories, meeting rooms, offices, and display areas. The lower story, which is underground, includes laboratories, a library, and mechanical equipment.

In plan view the building is U-shaped, with the opening of the U toward the east. The open area, in the central east part of building, is a courtyard that is at several levels. It is partly roofed over with collector panels, as explained below. The east end of the courtyard is open to the outdoors.

The roof of the building is of sawtooth form. There are six complete "teeth," 25 feet apart on centers, covering six bays. Each bay has several skylights. At the north end of the building there is a vestigial seventh tooth that supports an additional row of collector panels. Between successive teeth there are 4-foot-wide horizontal east-west walkways.

All of the lower story is underground, and large portions of the upper story are flanked by earth berms.

Energy conservation.   The building has excellent insulation and many energy-saving features. The above-grade outer walls, of 12-inch-thick concrete, are insulated on the exterior with 3½ inches of Styrofoam protected by a ¾-inch layer of cement plaster, giving an overall R-value of 16. The below-grade walls are insulated with 2 to 3

inches of Styrofoam and have an overall R-value of 23. The concrete floor of the building is insulated on the underside with 1 inch of Styrofoam. The window area is small, comprising 13 percent of the area of the above-ground exterior walls. Most of the window area is on the exposed south side of the upper story. All windows are double-glazed and are openable. All are fitted with manually operated sliding shutters employing 1 inch of urethane foam. Each of the building's twenty-eight skylights (likewise double-glazed) is equipped with a shutter consisting of a rotating insulating vane that is turned manually by means of ropes and pulleys. The shutter includes 2 inches of urethane foam. There is a heat-exchanger system for recovering heat from most of the air exhausted from the building. Rainwater is captured by the roofs and is used for watering lawns and flushing toilets and for fire prevention.

Collection There are seven arrays of water-type collector panels; one row covering each of the seven south-facing portions of the sawtooth roof. Each array is 110 by 7 or 8 feet and slopes 60 degrees. Two kinds of collector panels are used: Chamberlain Manufacturing Corporation panels and KTA Corporation panels.

The Chamberlain panels cover 4/7 of the collector area, specifically the four south rows. Each panel is 7 by 3 feet and employs two steel sheets that are formed and welded together to form a waffle-type pattern, with coolant flowing between the two sheets. The black coating used is selective. The panel is double-glazed with tempered glass. The panel backing includes 3 inches of fiberglass. The panels are connected hydraulically in parallel. The coolant consists of demineralized water and 50 percent propylene glycol. A flowrate of ⅓ to ½ gpm per panel is maintained whenever solar energy is being collected. The panels are supported by a steel framework.

The KTA panels cover 3/7 of the collector area, specifically the three north rows. Each panel is 8 feet high and 5 feet 3 inches wide and contains forty tube assemblies, 4 feet 10 inches long, lying along a horizontal east-west line. The assemblies are 2.12 inches apart on centers. Each assembly includes three concentric tubes: (1) a ½-inch-diameter copper tube that carries the coolant (demineralized water and 50 percent propylene glycol) and has a selective black coating (black chrome with $a/e = 0.95/0.07$), (2) a 7/8-inch-diameter tube of glass, and (3) a 2-inch-diameter tube of glass, the front half of which is clear; the back half is internally coated with silver and reflects radiation toward the copper tube. Both glass tubes are of soda-lime glass, with low iron content and high (95 percent) transmittance. The spaces between tubes contain air. Above the array of tube assemblies there is a 1/8-inch sheet of UVA Plexiglas which excludes snow, leaves, dust, etc. The two vertical manifolds are insulated with 4 inches of urethane foam. A coolant flowrate of 1 gpm per panel is maintained. A single 7½-hp centrifugal pump circulates coolant to both the Chamberlain and the KTA collectors.

The collection system as a whole is assisted by the white-shingled reverse slopes of the roof; these slopes, at 30 degrees from the horizontal, reflect some additional solar radiation to the collectors.

Heat is conveyed from the collection system to the storage system via heat exchangers.

Storage A total of 15,000 gallons of pure water is contained in two reinforced concrete tanks, of 10,000- and 5000-gallon capacities. The tanks are partially buried in the ground at the south side of the courtyard. Each tank is rectangular and 10 feet high. The below-ground portions of the tank walls are insulated on the exterior with 4 inches of foamglass and above-ground portions are insulated with 2 inches of urethane foam and 2 inches of Styrofoam. The tanks can be used identically, as a pair, or may be put to very different uses. A fan-coil system delivers hot air to the rooms via ducts. The distribution system was made extra large so as to be able to keep the rooms warm even when the temperature of the water fed to the coils is only 105°F, with the consequence that collection efficiency is especially high.

Auxiliary Heat Source At times when the solar heating system is unable to keep the rooms warm, as on many days in January and February, a heat-pump is used to extract heat from the cooler of the two storage tanks and deliver it to the distribution system (or to the hotter tank). Or the heat-pump can extract heat from 55°F well water. The heat-pump, which also can be used as a chiller, has a 30-ton cooling capacity. It was made by Durham-Bush Co. and is Model PCW-030

with oversize condenser. Much of the heat-pump operation is at night, i.e., at times of off-peak electric demand.

Domestic Hot Water About 70 square feet of the collector (i.e., about 1 percent) serves just the domestic hot water system. Final heating of the domestic hot water is by small electrical heaters situated close to the points of use.

Cooling in Summer During hot summer days the heat-pump is operated in reverse manner as a chiller. It extracts heat from the water that is circulated to the fan-coil system. The unwanted heat is discharged to the atmosphere by a cooling tower. The chilling may be applied either to the return water from fan-coil system coils or to chilled-water storage. At night, the building is cooled by forced circulation of outdoor air through the building. Note that little cooling is needed, inasmuch as (1) the building is largely underground, (2) exterior walls, etc. are well insulated, (3) south-facing windows are shaded by roof overhang and by vertical sunfins, and (4) the artificial illumination system, employing task lighting, contributes little heat.

Solar architect: Malcolm B. Wells. *Solar engineer:* Dubin-Bloome Associates. *General contractor:* John Lowry, Inc. *Cary Arboretum Project Manager:* Daniel F. Brown. *Owner:* New York Botanical Garden. *Cost of solar-heating system including design, construction, installation, modifications, supervision, and reporting:* $315,000. *Cost of solar heating data-acquisition-and-monitoring system:* $30,000. *Cost of building as a whole, including solar heating system:* $2,600,000. *Funding:* Mainly by Mary Flagler Charitable Trust. *Funding of solar heating system:* By ERDA, $122,000; by New York State Research and Development Authority, $76,000.

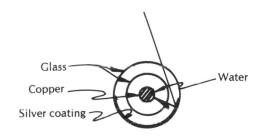

Cross section of KTA tube assembly

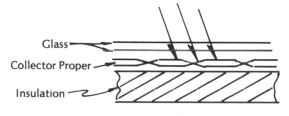

Cross section of Chamberlain collector

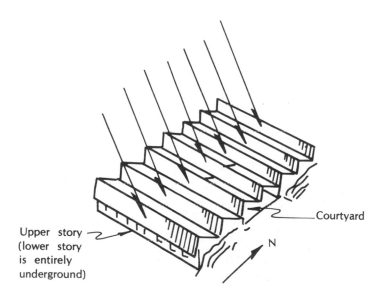

Courtyard

Upper story (lower story is entirely underground)

Quogue   41° N
(near E tip of
Long Island)
Dune Road, at South
Fork, facing Shinne-
cock Bay

Building: 2-story, 3000 sq. ft.
Collector: 510 sq. ft., tubular type
Storage: 1000 gals. water
% Solar-heated: 60

3000-sq.-ft. house, with living quarters on second and third stories, employing Owens-Illinois tubular-type collector

Building   This modern, two-bedroom, wood-frame house has living quarters on the second and third floors. The ground floor includes a carport, utility room, and studio. The second story, reached by an open staircase at the south, contains kitchen, living room, dining room, two bedrooms, and a central greenhouse area. The third story includes some additional living space and several balconies and sundecks. A dumbwaiter carries groceries, etc., from the ground floor to the second floor. There is no basement or attic. Most of the external surfaces slope 32½ or 57½ degrees. Walls and ceilings are insulated with 6 inches of fiberglass.

Collection   The 510-square-foot collector, mounted on a 57½-degree-sloping face at southwest part of building, faces straight south. With overall dimensions of 32 by 16 feet, the collector includes 384 Owens-Illinois, Inc., tube assemblies of tempered glass. Each assembly includes three concentric tubes: an outer tube 2 inches in diameter, an intermediate tube 1.6 inches in diameter, and an innermost tube ½-inch in diameter. The space between intermediate and outer tubes is evacuated to below 10^{-4} torr, to virtually eliminate heat loss

by conduction and convection to the outdoors. Both of the smaller tubes contain coolant, the inner tube serving as supply and the intermediate tube serving as return, and both join a massive, insulated, horizontal, east-west manifold containing supply and return headers. Each manifold serves two arrays of tubes: an array just above the manifold and an array just below it. Each array includes 24 tube assemblies, which are 4 inches apart on centers. Each tube is slightly less than 4 feet long, and the end farther from the manifold is fairly free; there are no coolant connections here; there is nothing to impede thermal expansion and contraction. The outer surfaces of the intermediate tubes are coated with a selective black coating, the a/e ratio of which is about 12. Between adjacent tube assemblies there is an open space, 2 inches wide, provided in order to reduce the tendency of a given assembly to shade its neighbors at times several hours before or after noon. Near noon, half of the incident direct radiation passes between the tubes and strikes a diffuse white backup surface there; a major fraction of the radiation reflected by that surface strikes the tube assemblies. In summary, nearly all of the incident radiation is absorbed by the tubes. The coolant is water, with no antifreeze or inhibitor. All of the tube assemblies are hydraulically in parallel. Arrays at very different heights above ground are served by different centrifugal pumps; in each array the pressure is about 15 to 20 psi.

Storage The 1000 gallons of water is stored in a 6-foot-diameter spherical fiberglass tank. Heat is distributed to the rooms by a fan-coil system.

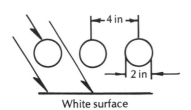

White surface

Cross section of portion of an array

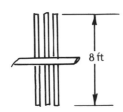

Portion of manifold, and 6 tubes

Water

Vacuum

Cross section of tube assembly

Auxiliary Heat Source   Two special fireplaces, with water circulating within the grates.

Domestic Hot Water   This is preheated by the solar heating system.

Cooling in Summer   None needed, none provided.

Problems and Modifications   The control system was found to be unnecessarily complicated and was simplified.

Solar engineering: Owens-Illinois, Inc. (J. Frissora et al.). *Architect:*   J. S. Whedbee. *Builder:*   Charles DeVoe and Sons. *Owner and occupant:*   Philip Barbash.

3000-sq.-ft. house with 1050-sq.-ft. water-type collector fabricated on-site

Fairview
(near Asheville)

Building Attached to the south side of this 54-foot-long, five-bed-room, wood-frame house is a 48-by-30-foot room containing a 40-foot-long swimming pool. Attached to the north side is a two-car garage. The walls contain 4 inches of fiberglass and the roof contains 6 to 8 inches. All windows are double-glazed. The house faces exactly south.

Building: 1½-story, 3000 sq. ft.
Collector: 1050 sq. ft., water type
Storage: 9000 gals. water
% Solar-heated: No estimate available

Collection The collector, on the main south roof, which slopes 50 degrees, is of special design and was fabricated on-site. The heart of the collector is a 1/8-inch aluminum plate which is screwed directly to the 2x8 roof rafters, thus eliminating the need for boards or plywood sheets. Copper tubes ¼ inch OD and 5 inches apart are affixed to the undersides of the aluminum plates by means of pop-riveted aluminum clamping channels and high-thermal-conduc-tivity epoxy cement. The glazing is double and consists of .040-inch Kalwall Sun-Lite (fiberglass and polyester). The insulating backing is of fiberglass. The coolant is water with bactericidal additive and with pH held between 7.0 and 8.0. No antifreeze is used; the system is drained automatically whenever the ⅓-hp centrifugal pump, con-trolled by a differential thermostat, stops. The water flows directly into the storage-system tanks; no heat exchanger is used. The roof of the swimming pool is covered with aluminum foil, which directs additional radiation to the collector.

Storage The 9000 gallons of water is contained in two 5000-gallon poured-concrete tanks situated beneath the living room floor. Each is rectangular and is 9 by 8½ feet by 8 feet high. The tank walls are 10 inches thick. The tanks are waterproofed internally with epoxy and are insulated with earth. The rooms are heated by a thermostatically controlled, central, hot-air circulation system that employs one ½-hp, 2000 cfm blower and a Trane coil that is supplied with hot water from storage.

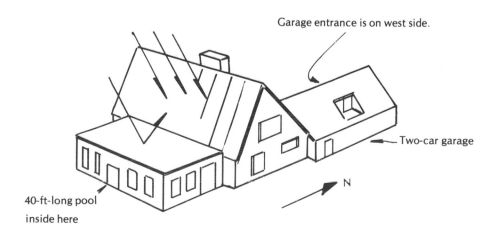

Garage entrance is on west side.

Two-car garage

N

40-ft-long pool
inside here

Auxiliary Heat Source   Two fireplaces, each with Heatilator. There is no furnace and no electric space-heating.

Domestic Hot Water   Preheated by the solar heating system.

Cooling in Summer   None.

Problems and Modifications   Because it is believed that the copper tubes of the collector may eventually become corroded by being repeatedly filled with water, then drained and filled with air, the owner is planning to (1) isolate the collector coolant from the storage-system water by means of a heat exchanger, (2) add anti-freeze to the coolant, and (3) keep the copper tubes filled with coolant at all times.

Solar engineer, designer, builder, owner, occupant: Robin M. Woodward.

Salisbury House

House 95% solar-heated by 640-sq.-ft. trickling-water-type collector

Salisbury   36°N
(50 mi SW of
Greensboro)
Southern Terrace, on
Roseman Road, off
U.S. 39 South

Building   This is a three-bedroom, 60-by-28-foot brick house with a built-in one-car garage at the west end. The 480-square-foot attic, which is unheated, was constructed economically of trusses that are made of 2x4's and are 2 feet apart. The thickness of fiberglass insulation is: walls, 3½ inches; attic, 6 inches. In general, this house is much like Thomason Solar House #6, but somewhat smaller and without a south porch, and with various special low-cost features.

Building: 1½-story, 1300 sq. ft.
Collector: 640 sq. ft., trickling-water type
Storage: 1600 gals. water and 25 tons of stones
% Solar-heated: 95

Collection   The trickling-water-type collector, on the steeply sloping south roof, is much like that of Thomason Solar House #6. The collector is single-glazed with single-strength glass.

Storage   The 1600 gallons of water is in a steel tank situated inside a bin containing 25 tons of 1½-inch-diameter stones. The general design is much like that of Thomason Solar House #6.

Auxiliary Heat Source   Oil-fired domestic hot water heater.

Domestic Hot Water   Preheated by passing through a small steel tank within the large tank mentioned above.

Cooling in Summer   A conventional air conditioner is operated throughout the coolest hours of the night to dry and cool air that is circulated through the bin-of-stones, cooling and drying them. During hot part of day, room air is circulated through the bin-of-stones and is cooled and dried by the stones.

Builder and owner: Dan Fisher Construction Company. *Supplier of solar equipment:* Carolina Solar Equipment Company, Inc. (B. D. McCubbins, D. L. Fisher, et al.). *Solar consultant:* H. E. Thomason. *Performance monitoring:* NASA, with funding by ERDA. *Cost of building:* About $39,000. *Cost of solar heating and cooling system:* $5600.

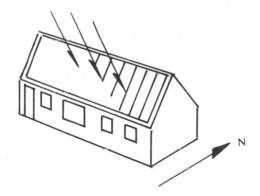

Statesville   35°N
(40 mi SW of
Winston-Salem)
Chipley Ford Road,
Rowan Acres

Building: 1-story, 1170 sq. ft.
Collector: 724 sq. ft., trickling-water type
Storage: 1250 gals. water and 24 tons
of stones
% Solar-heated: Near 100

Near-100% solar heating achieved with low-cost collectors mounted on top and end of carport

Building   This wood-frame and brick house has three bedrooms, attic, and full basement. The walls include 6 inches of fiberglass and the ceilings include 12 inches. The upper part of the basement also is insulated. The windows, with a total area of 100 square feet, are double-glazed and weatherstripped, and the outer doors likewise. The building faces east. At the south end of the building there is a two-car carport.

Collection   The two-part trickling-water-type collector is installed on the carport's roof and on its vertical south end. The trickling water is confined between two nesting corrugated sheets of .019-inch Reynolds aluminum with valleys 2.7 inches apart on centers. The two sheets are held apart by .019-inch aluminum spacer tabs. The wetted faces of the aluminum sheets have been zinc chromated, and, to fur-

ther reduce corrosion, the builder included a sacrificial getter in the hydraulic circuit. The coolant is water, with no antifreeze; the system drains automatically when the circulation pumps stop. Water is fed to each valley between the sheets by a ½-inch-diameter copper pipe that has one 1/32-inch-diameter hole per valley. The water flowing from the valleys is collected by a gutter. The upper edges of the corrugated aluminum sheets are joined by a shaped aluminum cap strip and the side edges are sealed with General Electric Company silicone cement. The collector supply and return pipes are enclosed in a single 4-inch-diameter plastic pipe and the remaining space in this pipe is filled with urea formaldehyde foam, foamed in place. The pumping system includes two pairs of 1/40-hp centrifugal pumps, which provide a total flowrate of 10 gpm. The collector is in two parts. The part on the slightly sloping carport roof is 24 by 22 feet and has an area of 524 square feet. The flowrate is about 5 gpm. In midwinter this part receives relatively little solar radiation because the sun is so low in the sky; therefore, the control system for this part starts the water circulating relatively late in the morning. If snow accumulates on this part, it may be quickly melted by distributing 57°F well water onto it via a perforated ¾-inch-diameter pipe. The other part, on the vertical south wall of the carport, is 24 by 8 feet and has an area of 200 square feet. In midwinter this part receives much solar radiation, and accordingly on sunny days the control system for this part starts the water circulating about one and a half hours earlier than the roof part starts operating.

Storage Water and stones are used. The 1250 gallons of water is contained in a two-piece prefabricated concrete septic tank (situated within the bin-of-stones described below) waterproofed with a variety of sealants, including Surewall fiberglass-reinforced mortar. The tank walls are 3 inches thick. Water is circulated from the bottom of the tank to the collector, and returns to the top of the tank. The 24 tons of 3-inch-diameter granitic stones is housed in a rectangular bin 16 by 8 feet by 7½ feet high. The bin is made of 8-inch-thick Zonolite-filled masonry lined with 2 inches of Styrofoam. The air plenum at the bottom of the bin is formed of hollow masonry blocks. The rooms are heated by forced circulation of air from the top of the bin; the air is driven through the main duct by a ⅓-hp blower and is distributed to individual rooms or sets of rooms by booster blowers of about 1/8 hp. Cold air returns to the bottom of the bin. The blowers are controlled by several thermostats.

Auxiliary Heat Source A 70,000-Btu/hr oil-fired hot water heater is used. It delivers heat to a serpentine coil in the upper part of the bin-of-stones and to a heat-exchanger coil in the main duct from the bin. There is also a high-performance Majestic fireplace that supplies scavenged heat to the bin.

Domestic Hot Water This is preheated by hot water en route from the collector to the storage system. Final heating is by the above-mentioned oil-fired heater. Note: In summer, only the near-horizontal part of the collector is operated, the other part is valved off.

Cooling in Summer   A conventional 10,500-Btu/hr air-conditioner is used in standard manner.

Problems and Modifications   Initially, in summer, the air conditioner was used to cool the bin-of-stones, and the bin was then used to cool the rooms. However, this was found to be less economical than conventional, direct use of the air conditioner.

Solar engineer, designer, builder: Solar Technology, Inc. (Nelson Brown). *Owner, occupant:*   Larry Woods. *Cost of building and solar heating system:*   About $30,000. *Cost of solar heating system alone:* $3000.

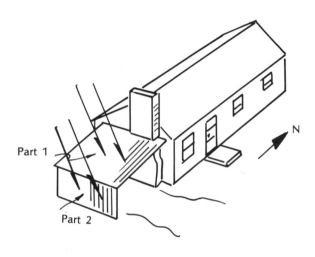

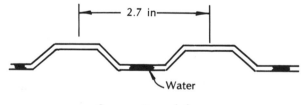

Cross section of the
two nesting sheets of
corrugated aluminum

House solar-heated by combination of trickling-water-type collector and a water-to-air heat-pump

Mansfield   40°N
(between Columbus
and Akron)
Marion Avenue
Road, near Twin
Lakes Golf Club

Building   This three-bedroom house has an attic and a crawl space but no basement. At the east end there is a built-in one-car garage. The building is 72 by 32 feet and faces exactly south. There is a covered terrace on the south side. The walls contain 3½ inches of fiberglass and the ceiling or roof contains 6 inches.

Collection   The trickling-water-type collector, on the south roof-wall, sloping 60 degrees, includes thirty-seven panels, all built on-site. The absorber sheet is of 28-gauge stainless steel with small ridges, or corrugations, that run up and down. The black coating is nonselective. The glazing, which is single, consists of Kalwall Premium Sun-Lite (fiberglass and polyester). There are twenty-four panels, each 8 by 2 feet, and thirteen panels 17 by 2 feet. The coolant is water with no antifreeze or inhibitor. It is fed to the tops of the panels by a 1-inch-diameter copper pipe equipped with a series of ¼-inch-

> **Building:** 1-story, 1800 sq. ft.
> **Collector:** 750 sq. ft., trickling-water type
> **Storage:** 8000 gals. water
> **% Solar-heated:** See text

diameter nozzles 4 inches apart. At the bottoms of the panels the water is collected by an insulated stainless steel gutter. The water is supplied to the collector by a 20-gpm Bell and Gossett centrifugal pump. In summer the panels are vented.

Storage The 8000 gallons of water is kept in two 4000-gallon concrete tanks (cisterns) under the garage. The tank sides are insulated externally with Styrofoam, and the tops are insulated internally. Thoroseal waterproofing is used. When the rooms need heat, heat is supplied by a water-to-air York Triton DW-30H heat-pump with 35,400-Btu/hr capacity and by a fan-coil system. The heat-pump takes energy only from the water in the storage tanks.

Percent Heated by combination of solar system and heat-pump: 62.

Auxiliary Heat Source Electric resistance heaters in ducts.

Domestic Hot Water This is not heated by the solar heating system.

Cooling in Summer The heat-pump takes heat from one tank and delivers it to the other, and water from the cooler tank cools the rooms by means of the fan-coil system. If the hotter tank were to become too hot, it could be drained and refilled with water from the city mains; but this has not been necessary and has not been done.

Problems and Modifications No significant problems have arisen.

Solar engineers, designers, architects, builders: Thomas Zaugg and John Zaugg. *Cost of building and solar-heating system:* About $80,000. *Funding:* Private.

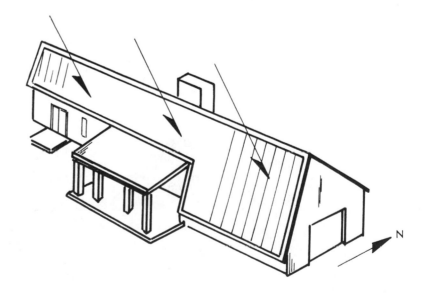

25,000-sq.-ft. library building 70% solar-heated by means of a 3260-sq.-ft. Owens-Illinois tubular collector

Troy 40°N
(60 mi W of
Columbus)
301 West Main Street

Building This includes the main story (16,000 square feet), partial basement (6600 square feet), and small attic. The walls are of concrete blocks and bricks. Walls are insulated with 3 inches of fiberglass and the roof has 6 inches. The respective R-values are 11 and 25. The windows are on the north and south sides only, with equal amounts on each and a total area of 1800 square feet. All windows are double-glazed.

Collection The collector consists of 102 panels of Owens-Illinois tube assemblies mounted on the south roof, which slopes 23 degrees. The panels themselves, mounted on A-frame supports of structural steel, slope 40 degrees. There are three rows of panels. Each panel is 8 by 4 feet and consists mainly of tube assemblies, each of which is 3½ feet long, employs three concentric tubes of glass, and has a vacuum jacket. The outside diameters of the tubes are 2.0 inch, 1.6 inch, and 0.6 inch. The outer annular space is a near vacuum (pressure less than 10^{-4} torr). The outer surface of the intermediate tube has a selective black coating with an a/e ratio of .9/.1. Each tube is perpendicular to an east-west line. Successive tube assemblies are 4 inches apart on centers, so that the spaces between assemblies are 2 inches wide. Behind the assemblies there is an array of concave aluminum reflectors.

Building: 1-story, 25,000 sq. ft.
Collector: 3260 sq. ft., tubular type
Storage: 5000 gals. water
% Solar-heated: 70 (predicted)

213

One end of each inner tube and each intermediate tube joins a horizontal manifold assembly, which carries input water and output water. The coolant is water with no antifreeze. Heat loss from the tube assembly is so small that even on typical nights in winter, no freezing will occur. On especially cold nights water from the (hot) storage tank is circulated through the tubes for a few minutes once per night.

Storage   The 5000 gallons of water is stored in a horizontal, cylindrical fiberglass tank buried in the ground near the west end of the building. The tank is insulated with 4 inches of urethane foam protected by a waterproof cover containing fiberglass. When the rooms need heat, hot water from the tank is circulated through the coils of a conventional air-handler. Or if hot water is currently being supplied by the collector, this water can be sent directly to those coils.

Auxiliary Heat Source   Existing electric heaters serving the forced-air system.

Domestic Hot Water   This is not heated by the solar heating system.

Cooling in Summer   A conventionally powered water-chiller is used. It may later be replaced with a solar-powered system.

Problems and Modifications   Early plans, later abandoned, called for installing collector assemblies on the horizontal roof also, and for providing a much larger storage system.

Solar engineering: Research Institute, University of Dayton (D. H. Whitford, et al.). *Manufacturer of solar equipment:*   Owens-Illinois, Inc. *Architect:*   Richard Levin and Associates. *General contractor:*   Starco, Inc. *General supervision, owner, user:*   Troy Library Board of Trustees. *Funding:*   In mid-1976 ERDA made a grant of $292,896 for the solar heating system.

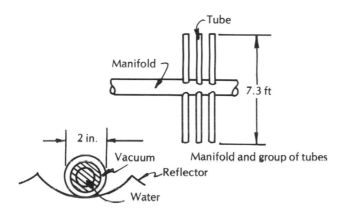

Cross section of tube assembly

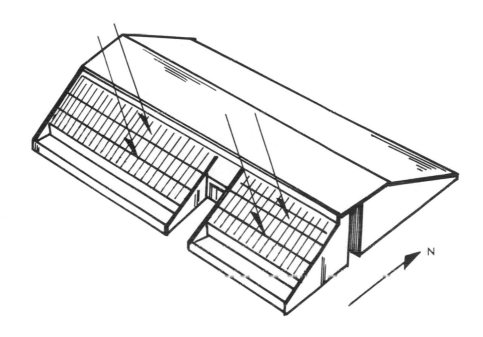

Wagoner   35°N
(in NE corner of
state)

Building: Four-level, 3700 sq. ft.
Collector: 760 sq. ft., water type
Storage: 2000 gals. water
% Solar-heated: 70

Four-level, four-bedroom house 70% solar-heated by 760-sq.-ft. water-type collector

Building   This four-level house, with four bedrooms, a loft, and a basement, has a floor area of 3700 square feet. There is also a two-car, attached garage. Excellent insulation is provided, and much use is made of earth berms. Windows and glass doors are double-glazed. The calculated heat loss when outdoor temperature is 0°F is 59,000 Btu/hr.

Collection   The collector, situated on the large south central roof, of conventional construction, that slopes 50 degrees, includes thirty-nine panels made by Energy Systems, Inc. Each is 6½ by 3 feet by 4½ inches thick. The absorber consists of a side-by-side array of extruded aluminum strips. At the center of each strip there is a specially formed recess, or channel, that firmly grips a copper tube in which the coolant flows. The black coating is nonselective. The panel is double-glazed with Thermopane of standard patio-door type. The panel backing includes 2½ inches of fiberglass and a sheet of tem-

pered Masonite. The coolant is water; no antifreeze is used. The panels are drained before freeze-up can occur. Heat is delivered to the storage system directly: no heat exchanger is used. Circulation is provided by a ⅓-hp Bell and Gossett pump; the maximum flowrate is 20 gpm.

Storage   The water is contained in a rectangular concrete tank in the basement. The basement walls form part of the tank. The inner face of the tank is insulated with 3 inches of Foamglas waterproofed with epoxy resin. When the rooms need heat, hot water from the tank (or from the electric boiler mentioned below) is circulated to a four-row fan-coil system.

Auxiliary Heat Source   20-kw electric boiler.

Domestic Hot Water   This is heated by a coil in the storage tank whenever that tank is adequately hot. Otherwise, heating is by an electric heater serving a special 50-gallon tank. About 95 percent of the water heating is solar.

Cooling in Summer   A standard compression-type air conditioner, with 4-ton capacity, is used.

Problems and Modifications   Some of the insulation of the storage tank deteriorated and was replaced. Plans have been made for installing a heat-pump to be used for cooling in summer and to assist the solar heating system in winter by extracting more heat from the storage system.

Solar engineer and architect: Alan Lower & Associates. *Owner and occupant:* Murray Engle. *Cost of building and solar heating system:* $80,000.

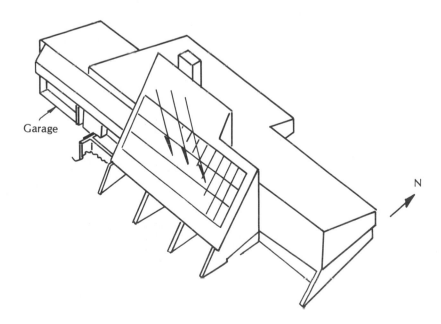

Garage

N

Coos Bay   43½ °N
(70 mi SW of Eugene)

Building: 1-story, 1650 sq. ft. **Collector:** 725 sq. ft., water type **Storage:** 8000 gals. water **% Solar-heated:** About 80

80% solar-heating achieved with two long, slender homemade collectors, a reflector made of aluminum foil, and a homemade 8000-gal. storage tank

Building This three-bedroom house has six rooms: living room, three bedrooms, kitchen, and laundry room. There is also a garage and greenhouse area at the east end. There is no attic or basement. The south and north roofs, each 19 feet wide, slope symmetrically 8 degrees from the horizontal. At the west end of the south roof there is a 7-foot overhang which, in summer, shades the living room and kitchen.

Collection There are two collector arrays, one on the roof and one on the ground. The combined area is 725 square feet.

The array on the roof is 80 feet long by 5 feet high. It runs along the ridge of the roof and slopes 82 degrees. The absorber is a corrugated, .017-inch sheet of aluminum with a nonselective black coating. The corrugations are 4 inches apart on centers and run east-west. Along each corrugation valley (except the lowest three) there is a horizontal ½-inch-diameter galvanized steel pipe which is tied snugly to the aluminum sheet with wire. The coolant is plain water. The glazing is single and consists of 30-by-20-inch sheets of glass 1/16 inch thick. Three such sheets, slightly overlapping, constitute one 5-foot-high glazing panel. The glazing is 1½ inches from the black aluminum sheet, which is backed by 1½ inches of fiberglass. Almost the entire south portion of the roof is covered by .001-inch household aluminum foil secured to the roof by a plastic roofing compound. The reflective foil increases the amount of radiation striking the collector proper by about 40 percent and increases the amount of energy delivered to storage by about 100 percent, according to estimates made by D. K. McDaniels, et al., and presented in *Solar Energy 17*, 277 (Nov. 1975). At each end of the collector array there is a near-vertical header, or feeder, pipe; thus the array of many parallel horizontal pipes is parallel-fed. Water from the upper end of the west header pipe flows by gravity convection to an insulated indoor 45-gallon surge tank and thence to the storage tank.

The array on the ground is slightly smaller (325 square feet, as compared to 400) but is of the same general design. It stands 65 feet north of the house.

Storage   The 8000-gallon water tank, of 1/12-inch steel, was welded together on-site by the owner-builder. It is situated beneath the west end of the building. Surrounding most of the tank there is a ½-to-1½-foot airspace. Ten inches of fiberglass insulation is used. Water from the bottom of the tank is pumped (by a ¼-hp centrifugal pump situated outdoors) to the east header pipe of the collector at a rate of 10 to 15 gpm. When the thermostat shows conditions to be unfavorable, the pump stops and all water returns to the tank. When the rooms need heat, hot air from the airspace around the tank flows, by gravity convection, to the rooms. The flow is regulated by thermostatically controlled dampers.

Auxiliary Heat Source   Electric baseboard heaters.

Domestic Hot Water   No information.

Cooling in Summer   None needed, none provided. The aluminized roof admits little heat. The 7-foot roof overhang helps exclude radiation. Dampers confine the heat in the storage tank.

Problems and Modifications   Initially there were some control problems, but several improvements made in the period 1973-1976 solved those problems. Originally there were horizontal pipes attached to the lower part of the rooftop collector as well as to the central and upper parts; but it was discovered that—even while the upper pipes were transporting much heat—the lower part of the collector was often so cool that the lowest pipes were transporting little or no heat. Accordingly these pipes were removed. After the rooftop collector had been in use for several years and had been found to perform well, the owner-builder added the second (ground-based) collector.

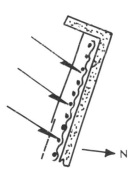

Vertical cross section of upper part of a collector array

Inventor, designer, builder, owner, occupant: Henry Mathew. *Performance studies:* by J. S. Reynolds, D. K. McDaniels, et al., from the University of Oregon and Oregon State University. *Cost of materials for the initial [1967] solar heating system:* $1000, according to the owner's estimate. Note that the owner himself fabricated and installed most of the components.

Boleyn House

Gladstone 45½°N
(10 mi S of Portland)
17610 Springhill
Place

Building: 2-story, 1800 sq. ft.
Collector: 430 sq. ft., water type
Storage: 3750 gals. water
% Solar-heated: 50

Utility-sponsored system employing set of three storage tanks connected in series to provide thermal segregation

Building This is a two-bedroom, wood-frame house that has an unfinished basement and no attic. The walls are insulated to R-11 and the roof to R-19. Of the 250-square-foot window area, 40 percent is on the south side. All of the windows are double-glazed. The house faces 6 degrees west of south.

Collection The collector, on a roof sloping 60 degrees, includes twenty-two panels (6½ by 3 feet) made by Revere Copper and Brass Company. These are in two rows of eleven each. The copper absorber sheet has a nonselective black coating. Rectangular-section copper tubes are attached to the sheet by means of clips and adhesive. The panel is double-glazed with tempered glass. The backing includes 2 inches of fiberglass. The coolant is water and 30 percent ethylene glycol. A centrifugal pump maintains a flowrate of 17 to 22 gpm. Flow is initiated whenever the Rho-Sigma differential thermostat finds the collector temperature to exceed the temperature of the coldest part of the storage system by a specified amount, typically 20 Fahrenheit degrees. Heat is delivered, via a heat exchanger, to the main storage system or may be delivered directly to an in-duct coil discussed below.

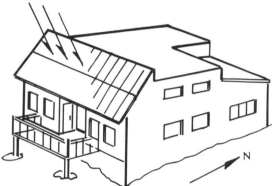

Storage The 3750 gallons of water is contained in three vertical 1250-gallon cylindrical tanks, of fiberglass and polyester, made by Hoffman Fiberglass Co. Each tank is 6 feet in diameter and 6½ feet high. The tanks are insulated and are arranged side by side in a small basement room that is insulated with 4 inches of fiberglass. The three tanks are connected in such manner as to permit a large degree of thermal segregation. The rooms are heated by a fan-coil system in the main duct. This system has three coils, A, B, C. Coil A is fed directly from liquid circulating from the collector. Coil B is fed from the main storage system. Coil C is fed from the auxiliary heating system tank discussed below (or, in coldest weather, from the electrical heater of that auxiliary system). During sunny days, coil A alone is used, ordinarily. When conditions are somewhat less favorable, coil B is used, and if it is not adequate coil C is used. A second centrifugal pump, providing a flowrate of 10 gpm, serves the loop that includes the main storage system, and a third pump, rated at 4½ gpm, serves the auxiliary system and coil C. In midwinter months the typical temperature of the storage system is so low (about 70 to 100°F) that collection efficiency is high.

Auxiliary Heat Source A 24-kw electric heater, operated at off-peak times, usually succeeds in keeping the 1250-gallon auxiliary heating system tank at adequately high temperature.

Domestic Hot Water This is preheated by the solar heating system. Solar heat provides 65 percent of the annual heat need.

Cooling in Summer None needed, none provided.

Instrumentation Extensive instrumentation for performance monitoring is provided.

Problems and Modifications A sensor of a differential thermostat corroded and malfunctioned. A pipe fitting of the collector was installed incorrectly and leaked. Some heat was lost from the storage system on cold nights as a result of gravity-convective circulation of water to the collector. Circulation of (untreated) water from collector to storage system was impeded by a buildup of algae or mineral deposit on screens. Consumption of electrical power by the solar heating system's four pumps and one fan is considered to be unnecessarily large.

Owner, occupant, system codesigner: Douglas Boleyn. *Architect:* R. K. Hansen and D. McClure. *Codesigner:* R. Josi of Portland General Electric Company. *Thermal engineering:* McGinnis Engineering, Incorporated, and Honeywell Incorporated. *Installation contractor:* Interstate Air-Conditioning and Heating. *General contractor:* Homecraft Construction Company. *Cost and funding:* The entire cost of the solar heating system and instrumentation, about $20,000, was funded by Portland General Electric Company.

Salem   45°N
1715 Altaview Drive
South

Building: 3-story, 1450 sq. ft.
Collector: 460 sq. ft., air type
Storage: 50-ton bin-of-stones
% Solar-heated: 50

House employs moderate-size air-type collector, built on-site, and large bin-of-stones.

Building   The three-bedroom, wood-frame house is situated on the steep west slope of a hill. Walls are insulated with 3½ inches of fiberglass and ceilings contain 6 inches of fiberglass.

Collection   The air-type collector, built on-site, is an integral part of the building's south face, sloping 60 degrees, and is 26 feet wide and 18 feet high. It employs corrugated aluminum sheets that are oriented so that the corrugations are horizontal. The black coating on the sheets is nonselective. The double-glazing consists of blemished 1/8-inch tempered glass sheets 76 by 34 inches. (The cost of blemished glass was 30¢ per square foot.) Air flows vertically in the 1-to-1½-inch space between glazing and black aluminum sheet. The air is driven by a ¾-hp blower. The collector is backed by 6 inches of urea formaldehyde foam.

Storage   The 50 tons (1200 cubic feet) of 2-to-3-inch-diameter stones is housed in a rectangular, 18-by-12-by-6-foot-high bin situated beneath the house. The bin is waterproofed and is insulated with 4 inches of rigid insulation. Within the bin, the air flows in a serpentine path defined by plywood baffles. When the rooms need heat, a 1/6-hp fan circulates room air through the bin.

Auxiliary Heat Source   Electric heater and forced-air system.

Domestic Hot Water   Preheated on flowing through a coil in the bin-of-stones.

Cooling in Summer   At night, the stones are cooled by circulation of cool night air. During a hot day, room air is circulated through the cool bin.

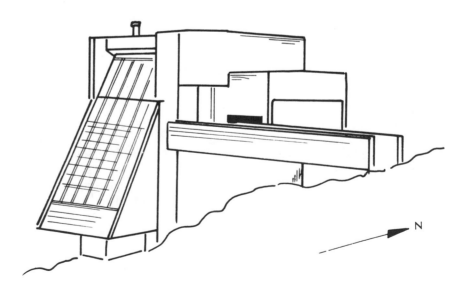

Problems and Modifications Often, the temperature of the bin-of-stones has been disappointingly low, presumably because (a) the insulation of the walls of the bin is not thick enough and (b) the start of heat-up of the bin has been delayed until well into the fall. Plans have been made for correcting these two matters. In midwinter there is significant shading of the collector by trees.

Solar engineer, architect, owner, occupant: William Bishoprick of Payne, Settecase, Smith and Partners. *Builder:* Entec Company. *Cost of solar heating system:* $5000. *Funding:* assisted by $3000 grant from Portland General Electric Company, which has been monitoring the system's performance.

Stoverstown   40°N
(30 mi S of
Harrisburg)

Building: 1½-story, 1250 sq. ft.
Collector: 520 sq. ft., air type
Storage: None, other than the house itself
% Solar-heated: 50

50% solar heating provided by homemade vertical air-type collector and no storage system

Building   This is a three-bedroom dwelling with no basement. The second story, a half-story, supports the collector and serves as an attic; it is not heated. The first story is 65 by 20 feet and has an area of 1250 square feet. An 18-inch-wide visor, or canopy, shades the first-story windows in summer. The first-story south-facing windows are double-glazed and have an area of 80 square feet. Note: The house and the solar collector were somewhat patterned on a solar-heated house (Peabody House) built in Dover, Massachusetts, in 1948 by Maria Telkes and others, except that Lefever House has no storage system.

Collection   The entire south wall of the second story, 65 by 8 feet, is double-glazed and serves as the collection area. The glazing consists of glass sheets 7/32 inch thick. In all, there are fifteen panels, each 8 by 4 feet. Radiation passing through the glass strikes a nonselective black surface, and fan-driven air carries heat from that surface, via ducts, to several 4-foot-wide closets (bins) between rooms, and thence to the rooms. The system employs one 1/6-hp blower and two 20-inch-diameter, 800-rpm fans.

Storage   None, other than the building itself. The builder had intended to install phase-change storage material in the above-mentioned closets; but no such material has been installed.

Auxiliary Heat Source   Gas-fired furnace with forced-air system. Also a Franklin stove.

Domestic Hot Water   No information.

Cooling in Summer   None. The above-mentioned visor blocks most of the radiation approaching the first-story south windows. Roof overhang protects the vertical collector.

Problems and Modifications   No significant problems have arisen in twenty years of operation. Four of the inner glazing sheets cracked, but replacement seemed unnecessary. The insulation of walls and ceiling has been improved by the addition, to certain areas, of aluminum-foil insulation and also Rapcofoam. A Franklin stove was installed.

Designer, builder, owner, occupant: Harold R. Lefever of Sonnewald Service.

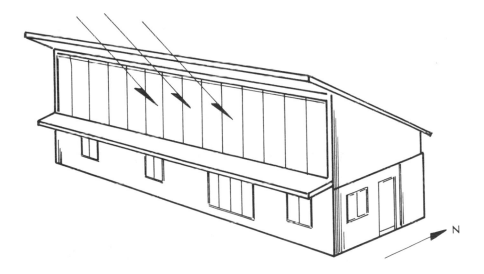

225

Jamestown, 41½° N
Lawn Ave.

Building: 1-story, 960 sq. ft.
Collector: 360 sq. ft., water type
Storage: Shallow, 138-ton, stones-and-water bin
% Solar-heated: 100

100% solar-heated by water-type collector and shallow stones-and-water bin that extends under entire house

Building This one-story, two-and-a-half-bedroom, wood-frame house is 64 feet long, including built-in two-car garage, and is 24 feet wide. There is a small, unheated attic and no basement. Heated floor area is 960 square feet. The house faces 10 degrees west of south. On the east, north, and west sides there are earth berms and few windows. The 12-inch-thick concrete walls are insulated with 6 inches of fiberglass and the ceilings are insulated with 12 inches of fiberglass. The windows are double-glazed.

Collector The 360-square-foot water-type collector, which is 60 by 6 feet, is mounted on the roof, which slope 57 degrees. The fifteen collector panels, each 6 by 4 feet, were made by Solar Homes, Inc. The absorber is a 16-ounce sheet of copper to which ½-inch-diameter copper tubes have been soldered 6 inches apart on centers. To lessen the stress that the tubes impose on the sheet when temperature differences occur, the fabricator employed, instead of one single large sheet, four smaller side-by-side sheets that are slightly and loosely overlapped. The black coating is a heat-resistant Rustoleum paint and is nonselective. The glazing is double and consists of Kalwall Sun-Lite .040 inch thick. The panel backing includes 6 inches of fiberglass. The water contains no antifreeze; it is drained before freeze-up can occur. The water is circulated to the collector by a 1/12-hp pump. The water flows directly into the storage system: there is no heat exchanger.

Storage   Heat is stored in 13 tons of water, 65 tons of one-inch-diameter stones, and 60 tons of concrete (of bin walls, etc.), with a total thermal capacity of 70,000 Btu/°F. The bin, of poured concrete, is 38 by 22 feet and 16 inches high. The water and stones are commingled in it. The stones serve the important additional purpose of supporting the large-area top of the bin. The top of the bin consists of a 4-inch-thick concrete slab and the bottom is a 6-inch slab; these two together weigh 60 tons. Under the bin there is a 4-inch insulating layer resting on another 6-inch slab; the insulating layer consists of sheets of Styro SM with interleaved layers of asphalt and polyethylene. The bin underlies the entire living area of the house. Between the top of the bin and the floor there is a 2-foot airspace, or plenum, that receives heat from the bin. At all times during the winter some heat flows upward from the tank and two-foot airspace, through the carpeted wooden floor, into the rooms. When additional heat is needed, registers in the room are opened and hot air in the two-foot airspace is allowed to circulate into the rooms, via ducts, by gravity convection; or, on rare occasions, a thermostatically controlled blower is operated to greatly speed the circulation.

Auxiliary Heat   None. There is no woodstove, no fireplace, no chimney.

Domestic Hot Water.   Preheated in coil in bin.

Cooling in Summer   None.

Problems and Modifications   At the outset, as a precaution, a water-to-air heat-pump was installed, it being thought that, in mid-winter, the storage bin might sometimes become too cold to heat the house, and that the heat-pump would then be needed to extract additional heat from the bin. In fact, the bin seldom became colder that 90° or 100°F and there was no need to operate the heat-pump, which was subsequently disconnected.

Designer and builder: Spencer Dickinson. *Manufacturer of solar heating system:*   Solar Homes, Inc. *Owner and occupant:*   Tobey Richards. *Funding:*   Private. *Cost of entire building, including solar heating system [but not land]:*   About $40,000. *Cost of solar heating system:* $8000. *Cost of operating the solar heating system:* $2 per month.

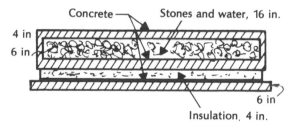

Cross section of storage system
(not drawn to scale)

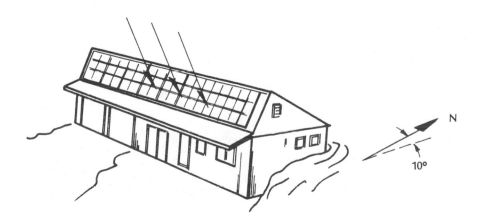

Little Compton 41½ °N
(in SE corner of state)
West Main Road

Building: 2-story, 1500 sq. ft. addition
Collector: ⎱ Combination passive and
Storage: ⎰ active systems; see text
% Solar-heated: About 90

A 1500-sq.-ft. addition and two existing rooms are 90% solar-heated by combination of greenhouse passive system and air-type active system

Building The two-story, 1500-square-foot solar-heated addition, built in 1975, is attached to a larger structure that was built in 1968 and is largely electrically heated. The new structure includes the living-dining room, one bedroom, a shop, mechanical room, and greenhouse. The floor of the lower-story living area is of concrete. The walls are insulated externally with 3½ inches of fiberglass batts and 1 inch of Styrofoam. The roof is insulated with 6½ inches of fiberglass, and the concrete basement walls are insulated externally with 3 inches of sprayed urethane foam. The very large south window area is described below. The window area on the east side is small and the area on the west side is large. All windows serving the living area are double-glazed; Thermopane and Anderson windows are used. The building faces exactly south.

Solar-Heating System Two solar-heating systems are used; one passive and the other an active air-type system.

Passive Solar Heating System. Radiation enters by the 60-degree-sloping window-wall of the greenhouse. This wall, which faces exactly south, is 30 feet long by 8 feet wide, and all of its area of 240 square feet is single-glazed. The greenhouse floor, which is 32 feet long by 10 feet wide and is 2 feet lower than the living-room floor area, is covered with a 1-inch layer of pebbles. The window-wall is insulated at night by means of sixteen automatically operated, solar-

powered Skylids, made by Zomeworks Corporation. There are four tiers of Skylids mounted one above another; the Skylid axes are horizontal, running east-west. When closed, the foam-filled Skylids provide a thermal resistance of about R-5. The walls, floor, and potted plants of the greenhouse absorb much solar energy. If the sliding Thermopane doors between greenhouse and living area are open, hot air from the greenhouse may circulate into the living area by gravity convection; otherwise, a blower may circulate this hot air through a small bin-of-stones beneath the greenhouse floor.

Active Solar Heating System.   There is a 360-square-foot array of air-type collector panels on the upper part of the 60-degree-sloping south wall. The eighteen panels, made by Sunworks, Inc., are arranged in two rows of nine panels each. Each panel is 7 by 3 feet. The heart of the panel is a finned sheet of copper with a selective black coating; the a/e ratio is .9/.1. The fins are on the back of the sheet, and air flows in the space between fins. Air is driven through the collector array by a two-speed, ½-hp blower. The panels are single-glazed with 3/16-inch tempered glass. Heat is stored in a 1050-cubic-foot, 50-ton bin-of-stones in the lower east part of the new structure. The stones are about 4 inches in diameter. The bin, of poured concrete, is 18 by 10 feet and 8 feet high. It is insulated on the inside with 4 inches of sprayed urethane foam protected by Sheetrock. There are 1-foot-high plenums at top and bottom of the bin. The direction of airflow within the bin is downward irrespective of whether the incoming air is from the collector or from the rooms. The control system includes five motorized dampers. Three modes of operation are available: collector heat to storage, collector heat to rooms, stored

heat to rooms. It has been found that there is seldom a need for the heating mode in which hot air from the collector is circulated directly to the rooms: whenever the sun is shining and the collector is operating, the greenhouse also is collecting much heat which may flow to the living area by gravity convection.

The bedroom and living-dining room in the new structure and the two south rooms of the old structure are heated by the active system. The two solar heating systems provide about 90 percent of the heat needed by 2000 square feet of old and new living and greenhouse areas in the two structures.

Auxiliary Heat Source   Jøtul wood-burning stove.

Domestic Hot Water   Just east of the new structure there are three Sunworks water-type panels. Each is 7 by 3 feet, slopes 45 degrees, and is made mainly of copper. These panels serve a 65-gallon tank that is situated in the upper southeast part of the new structure and is at higher level than the panels. The flow of water is by gravity convection. The preheated water in the 65-gallon tank flows to two existing 75-gallon domestic-hot-water tanks that contain electrical heating elements.

Cooling in Summer   There is no formal cooling system. The greenhouse and the air-type collector are vented.

Problems and Modifications   Initially the blowers did not operate at optimum speed. This was corrected by changes in the belt-drive systems.

Solar architect: Travis Price. *Solar engineering consultant:* Everett Barber. *Owner and occupant:* Junius Eddy. *Cost of solar equipment and pertinent labor:* $6000 to $8000.

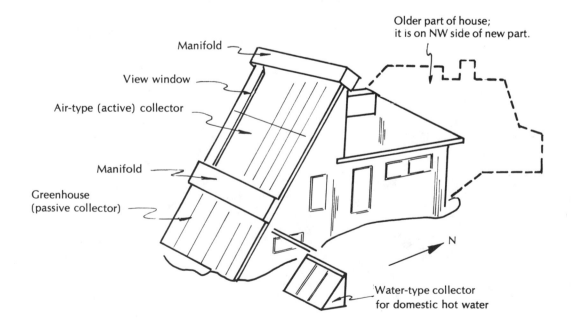

Manifold

View window

Air-type (active) collector

Manifold

Greenhouse
(passive collector)

Older part of house;
it is on NW side of new part.

N

Water-type collector
for domestic hot water

Carpenter House

100% heated by solar-and-heat-pump system

South Kingston   41½°N
Pond Street

Building   This one-story, two-bedroom, wood-frame house is 64 feet long, including built-in one-car garage, and is 24 feet wide. There is also a finished and heated walk-out basement; its area is 1200 square feet. The total heated area is 2100 square feet. The attic is unheated. There is 6 inches of fiberglass insulation in the walls and 12 inches in the ceiling. The basement floor, a concrete slab, rests on a 2-inch layer of Styro-SM board. All windows are double-glazed. The building faces exactly south.

Collector   The 360-square-foot, water-type collector, 60 by 6 feet, is mounted on the roof, which slopes 57 degrees. The fifteen collector panels, each 6 by 4 feet, were made by Solar Homes, Inc. The absorber is a 16-ounce sheet of copper to which ½-inch-diameter copper tubes, 6 inches apart on centers, have been soldered. To minimize the stress of the tubes on the sheet when temperature differences occur, the fabricator employed—instead of one single large sheet—four smaller side-by-side sheets that are slightly and loosely overlapped. The black coating, a heat-resistant Rustoleum paint, is nonselective. Double glazing, of .040-inch Kalwall Sun-Lite, is used. The backing includes 6 inches of fiberglass. The water, which contains no antifreeze, is circulated by a 1/6-hp pump and is drained be-

Building: 1-story, 1200 sq. ft.	
Collector: 360 sq. ft., water type	
Storage: 5000 gals. water	
% Solar-heated: See text	

fore freeze-up can occur. The heat is delivered directly to the storage system: no heat exchanger is used.

Storage   The 5000-gallon water tank, situated in the northeast corner of the basement, is of poured concrete. It is rectangular and is 20 by 8 feet and 6 feet high. It is waterproofed on the inside with three coats of PPG epoxy. The tanks walls are insulated on the outside with 4 inches of fiberglass, and the tank rests on a 4-inch slab of load-bearing Styro-SM board. Heat is extracted from the uppermost (hottest) part of the tank by a Vaughn (Wescorp) Solargy EC-44 water-to-air heat-pump, rated at 44,000 Btu/hr, and is delivered to the rooms by a forced-air system. The typical input and output temperatures of the heat-pump are 55 to 65°F and about 160°F, respectively. The typical value of COP of the heat-pump is 3 (predicted) or about 4 (predicted) if heat from heat-pump motor, etc., is included. (Note: The sole source of heat for the heat-pump in winter is the above-mentioned tank of water.)

Percent Solar Heated   By combination of solar system and heat-pump: 100. Percent attributable to solar system: 75.

Auxiliary Heat Source   None.

Domestic Hot Water   This is preheated in a coil in the above-mentioned tank. Final heating—both in winter and in summer—is by the above-mentioned heat-pump.

Cooling in Summer   The heat-pump is operated in reverse manner to extract heat from room air. The rejected heat is supplied to the domestic hot water system or to the above-mentioned tank.

Problems and Modifications   Initially, backup electrical heating was provided. However, no need for such heating arose, and the system was dismantled.

Designer and builder: Spencer Dickinson. *Manufacturer of solar heating system:* Solar Homes, Inc. *Owner and occupant:*   Philip Carpenter. *Funding:*   Private. *Cost, not including land:* $39,000. *Incremental cost of solar heating system:* $8000. *Cost to operate the heat-pump-and-solar system per winter:* $140.

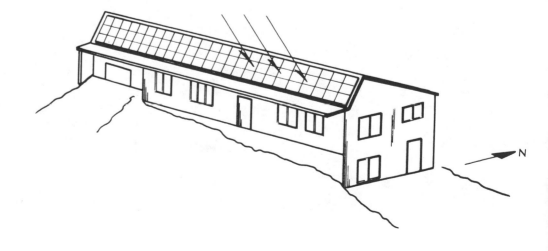

House 80% solar-heated by trickling-water-type collector and a 2000-gal. storage tank within a 50-ton bin-of-stones

Woonsocket 41½°N
55 Marian Lane

Building This 44-by-26-foot, high-peak, Cape-type house is of wood-frame construction with a brick south face. It has a full basement and no garage. The second story is not finished, not insulated, and not heated. Typical thickness of insulation in the living space is: wall, 4½ inches; ceiling, 3½ inches. The fifteen windows are of modest size and are double-glazed. The house faces exactly south.

Collection The collector is of trickling-water type and occupies a 43-by-19½-foot area on the south roof, sloping 52 degrees. The coolant (rainwater collected from the roof, with no antifreeze or inhibitor added) is fed by a ⅓-hp centrifugal pump to each valley of the .026-inch-thick sheet of corrugated aluminum via a ½-inch-diameter copper pipe that has one 1/16-inch-diameter outlet hole per valley. At the base of the collector the water is collected by an enclosed gutter of .032-inch aluminum and flows directly into the water tank of the storage system. The water drains from the collector automatically when the pump stops. The collector is single-glazed with 4-by-2-foot sheets of single-weight glass. The edges of the panes are sealed with neoprene strips that have a U-shaped cross section.

> **Building:** 1½-story, 2200 sq. ft.
> **Collector:** 830 sq. ft., water type
> **Storage:** 50 tons of stones, 2000 gals. water
> **% Solar-heated:** 80

Storage   The 4-to-8-inch-diameter stones are contained in a rectangular poured-concrete bin in the basement. The bin is 24 by 14 feet by 7 feet high. The sides, top, and bottom of the bin are insulated with 2 inches of Styrofoam. In the central upper region of the bin there is a horizontal, 5-foot-diameter cylindrical steel tank that contains 2000 gallons of water. Water from the bottom of the tank is circulated to the collector; the return is to the top of the tank. When the rooms need heat, a ⅓-hp blower circulates room air through the bin-of-stones.

Auxiliary Heat Source   This consists of the domestic hot water heater described below.

Domestic Hot Water   This is preheated by a copper coil within the upper part of the above-mentioned steel tank. The coil is made from a 150-foot length of ¾-inch-diameter copper tube. Final heating is by an oil-fired domestic hot water heater employing a 50-gallon tank maintained at 140°F.

Cooling in Summer   None.

Architect: R. E. Grenier. *Builder and owner:* R. L. G. Associates, Inc. *Construction and solar-system fabrication:* R. A. Montanari. *Cost of building and solar heating system:* About $40,000. *Cost of solar heating system:* $4000.

Attic itself is air-type collector, serving shallow bin-of-stones extending under entire floor area

Greenville   35°N
(in NW corner of
state)
At corner of Summit
and Pinehurst Streets

Building This is a three-bedroom, wood-frame house with crawl space, small attic space, and built-in one-car garage. The window area is small, and windows are double-glazed. The walls are insulated with 3½ inches of fiberglass and 1 inch of Styrofoam. Ceilings are insulated with 12 inches of fiberglass.

Building: 1-story, 1024 sq. ft.
Collector: 371 sq. ft., air type
Storage: 40-ton bin-of-stones
% Solar-heated: 70

Collection The collector consists of the south roof and attic. The south roof, 50 by 7½ feet, slopes 50 degrees and is transparent, allowing solar radiation to penetrate deep into the attic. The glazing is double and is of Filon corrugated fiberglass-filled plastic. Radiation that passes through the glazing travels 1 to 10 feet and strikes horizontal sheets of plywood, with nonselective black coating, that form the floor of the attic. The volume of space between glazing and black plywood is 3500 cubic feet; when the sun is shining, much hot air accumulates in this space. Air is circulated to and from this space at 1000 cfm by a ⅓-hp blower situated in a closet; the air is delivered to the rooms or to the storage system.

Storage The 40 tons of 1½-inch-diameter stones (railroad ballast) is in a shallow bin that extends beneath the entire floor area, occupying

almost all of the crawl space. The bin is 40 by 26 feet by 1 foot high. It is insulated on the bottom and sides with 1 to 2 inches of Styrofoam. Above the bin there is a 4-inch layer of fiberglass and directly above this there is a "floating" wooden floor, supported by the mass of stones. Hot air from the collector enters the bin via a central horizontal east-west channel, or trench, that is 1 foot deep. The air then flows horizontally northward or southward to parallel channels along north and south edges of the bin, just inside the foundation walls. When the rooms need heat, air is driven through the stones in the reverse direction and flows into the rooms. The flow is controlled by dampers. There are eight operating modes.

Auxiliary Heat Source   Gas-fired domestic hot water heater that serves special coils in the ducts that carry air from bin to rooms.

Domestic Hot Water   This is preheated by a special absorber plate that rests on the floor of the attic. Inasmuch as this space is insulated, and the winters in South Carolina are not very cold, no antifreeze is needed.

Cooling in Summer   The bin-of-stones is cooled at night by circulation of cool night air. During hot days room air is circulated through the bin.

Problems and Modifications   None. (Note: In 1977 a second house of this general type was built. This house, in Athens, Georgia, is slightly superior in some ways. For example, the outer glazing is of tempered glass and the inner glazing is of .005-inch mylar. Plans for this house [Plan 7220] are available from any USDA Agricultural Research Service office.)

Solar engineering and general design: U.S. Department of Agriculture, Agricultural Research Service, Rural Housing Research Unit (esp. H. F. Zornig and L. C. Godbey). *Builder, owner, occupant:*   Mike Granger of Helio-Thermic, Inc. *Funding:*   Private. *Cost:*   About $25,000.

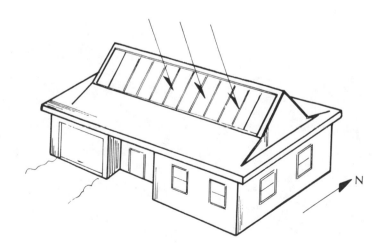

Ponton House

3500-sq.-ft. adobe house and 1500-sq.-ft. swimming-pool-and-green-house passively solar-heated by large ground-floor window area and rooftop monitors

Alpine
(alt. 5200 ft.; a 2000-degree-day site)
(in SW part of state)

Building   A one-story, two-bedroom, 5000-square-foot house, of which 1500 square feet consists of a swimming-pool-and-greenhouse. The walls of the house are of 20-inch-thick adobe, with no added insulation. The floor consists of bricks resting on 2 inches of urethane foam, with sand below. The ceiling includes 2 inches of urethane foam. All ground-floor windows are of Anderson double-glazed type.

Passive Solar Heating System   Three kinds of window areas contribute to solar heating:

(a) 630 square feet of single-glazed glass windows, which slope 55 degrees, mounted on monitors on the horizontal roof. Each such window is backed up by a set of Zomeworks Corporation automatic Skylids which open when the sun shines and close when the sky becomes heavily overcast or is dark. The combination of window and closed Skylid has an R-value of about 5.

Building:	1-story, very large area
Collector:	Passive; see text
Storage:	
% Solar-heated:	90

(b) 225 square feet of Anderson double-glazed windows, sloping 55 degrees. These serve the pool-and-greenhouse.

(c) 100 square feet of vertical Anderson double-glazed windows serving the bedrooms.

Energy is stored in 360 tons of walls and floors and in 80 tons of water in the pool. The pool remains at 85° to 90°F throughout the winter.

Auxiliary Heat Source   Two fireplaces, used mainly for esthetic effect.

Domestic Hot Water   Heated by Zomeworks thermosiphon solar system employing an 80-square-foot, single-glazed collector.

Cooling in Summer   None. Little cooling is needed because the adobe walls provide much insulation, the walls and floor have large thermal capacity, eaves shade the windows serving the pool-and-

greenhouse, and the Skylids are tied shut to prevent radiation from entering the rooms via the monitors. The pool remains at about 72°F throughout most of the summer.

Solar engineer: Sun Mountain Design; also Zomeworks Corporation. *Architect:* David Wright. *Owner and occupant:* A. R. Ponton.

Swimming pool & greenhouse region is beneath these monitors

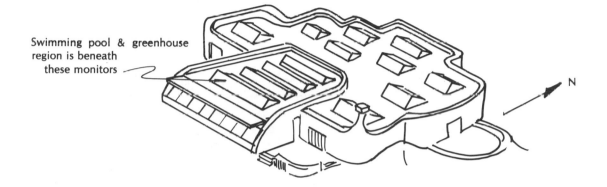

N

UTA Discovery 76 House

Arlington   33°N
(suburb of Dallas)
On campus of
University of Texas at
Arlington

Building: 1-story, 1900 sq. ft.
Collector: 420 sq. ft. of concentrating
               collector boxes
Storage: 3000 gals. water
% Solar-heated: 75

75% solar heating and 60% solar cooling provided by 42 Northrup tracking, concentrating collector boxes

Building This is a three-bedroom house, with a breezeway and two-car garage at the northwest corner and no attic or basement. The building was used initially as a laboratory. In constructing the house, much use was made of a new kind of material called Tectum II, which provides structural strength, thermal insulation, and absorption of sound. Made by National Gypsum Company, it is composed of wood fibers and also incorporates 1½ inches of urethane foam. The aggregate area of windows in the central parts of the south and north walls is 200 square feet, while the area of east and west windows is negligible. All windows are 1-inch-thick assemblies of two sheets of Libby-Owens-Ford ¼-inch glass with a hermetically sealed airspace between them.

Collection The 420 square feet of collector is distributed in two linear arrays, one 11 feet north of the other. These are mounted on the horizontal roof of the central part of the building. Each array is 47 feet in east-west dimension and 10 feet in north-south dimension. Each array includes twenty-one Northrup collector boxes, each of which is 10 feet long, 1 foot wide and 1 foot deep. The boxes are 26 inches apart on centers, and each is tilted with its north end higher than the south end, the typical slope being 27 degrees. The top of

each box is a 10-foot-long, 12-inch-wide cylindrical plastic fresnel lens with 12-inch focal length. About 12 inches below the lens is the absorber proper: a 10-foot-long, 2-inch-wide flattened copper tube with selective black coating. Coolant (water and antifreeze) enters the tube at one end and leaves by the other. The boxes are slowly turned to track the sun; they rotate about the longitudinal axis and are linked together mechanically so that one tracking motor can control many boxes. The total angular range of tracking is 140 degrees.

Storage The 3000 gallons of water is stored in three 1250-gallon tanks situated in the garage. Each tank is a vertical cylindrical tank of fiberglass and plastic and is insulated. The group of three tanks is enclosed in an insulated housing. Heat is delivered to the rooms by a forced-air system.

Auxiliary Heat Source Lennox Industries two-speed water-to-water heat-pump. Also an electric furnace and forced-air system.

Domestic Hot Water Preheated by the solar heating system. Final heating is electric.

Cooling in Summer Use is made of a 3-ton Arkla-Servel lithium-bromide absorption cooling system, powered by 200°F water from the solar heating system. The solar heating system is expected to provide about 40 percent of the power needed for cooling.

Problems and Modifications Initially the solar-powered cooling system was inoperative because of faults associated with the vacuum pump used. Repairs have been made. Some of the first collector boxes were not fully weatherproofed; the difficulty was soon corrected. An insulated housing was provided for the three storage tanks because the insulation on those tanks was not adequate.

Architect and solar engineer: Todd Hamilton of the School of Architecture of the University of Texas at Arlington. *Initial manager of project:* G. W. Lowery of the Department of Mechanical Engineering. *Subsequent manager of project:* T. J. Lawley. *Owner:* University of Texas at Arlington. *Funding:* By Texas Electric Service Company.

A collector box

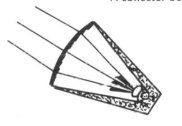

Cross section of a collector box (not to scale)

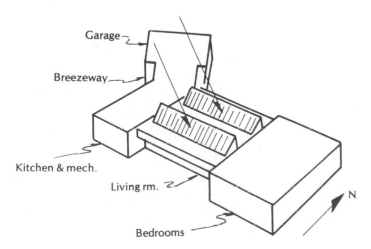

Wilder House

Hartford   43½°N
(near Hanover, N.H.)
On Route 5, north of
Wilder Village

Building: 2-story, 1500 sq. ft.
Collector: 576 sq. ft., air type
Storage: 40-ton bin-of-stones
% **Solar-heated:** 70

Two air-type collectors and 40 tons of stones provide 70% solar-heating

Building   This is a two-bedroom, 32-by-24-foot house with sun deck, entry deck, and an attached two-car garage. The living room has a cathedral ceiling, and the dining room is cantilevered. Most of the rooms are on the second story. The first story, partly recessed into the earth, contains two bedrooms and bath. The foundations are of wood, pressure-treated with preservative. Walls and ceilings are insulated to R-40 with a combination of fiberglass and three kinds of organic foam: Styrofoam, urethane, and urea formaldehyde. Extensive precautions have been taken against moisture deposition and air leakage. The windows have an area of 250 square feet and are double-glazed. The heat demand of the house is 8500 Btu per Fahrenheit degree-day. The house faces 46 degrees west of south.

Collection   The vertical air-type collector consists of five large panels, specially designed and built on-site. There are three 12-by-8-foot panels on the house proper and two 12-by-12-foot panels on the garage. The heart of a panel is a nonselective black coating on a "stucco-embossed" sheet of .040-inch aluminum that has been formed into a ribbed surface with 4-inch-wide, 1-inch-deep channels. The glazing is double, consisting of two layers of .040-inch Kalwall Premium Sun-Lite. The backing includes foil-faced CDX plywood and a 4½-inch-thick combination of Styrofoam and fiberglass. Air flows horizontally in the .66-inch space behind the black ribbed sheet, between it and the foil face of the plywood. The air in the space between the glazing and the black absorbing surface is virtually stagnant; the space is sealed except for small weep-holes that permit escape of any moisture that might be present here. The air passes through the panels in parallel.

Storage   The 1½-to-2-inch-diameter stones are contained in a rectangular bin, 12 by 12 feet by 8 feet high, situated in the northwest corner of the basement. The bin is of fir plywood and 2x6-inch spruce studs and is insulated to R-30. Distribution channels at the base of the bin are provided by spaced rows of concrete blocks. A partition divides the bin into two side-by-side halves and the air flows through the two halves in series, in effect doubling the pathlength and increasing the degree of thermal stratification achieved. Just above the 5-foot-high stone filling there is a 2-foot-high plenum. Airflow to and from the collector is maintained by a ½-hp blower. When the rooms need heat, room air is circulated through the bin by a ⅓-hp blower and/or through the plenum of the auxiliary heat source. Control is accomplished by motorized dampers. If, on sunny days in winter, the south room with clerestory overheats, warm air from the uppermost part of the clerestory is automatically circulated, by a central distribution fan, to cooler rooms of the upper and lower stories. There is also some passive solar heating via the southeast and southwest view windows. In the especially cold winter of 1976-77 the solar heating system provided 60 percent of the winter's heat need. It is expected to provide about 70 percent in a typical winter.

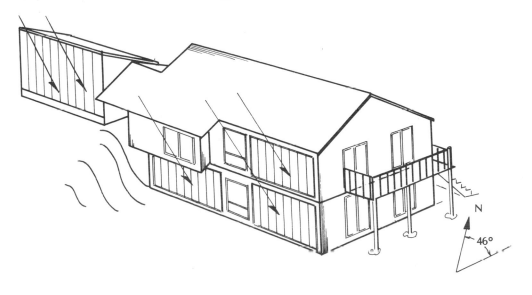

Auxiliary Heat Source A 10-kw electric furnace. Also a 25,000 Btu/hr cast-iron wood-burning stove with special air supply from outdoors.

Domestic Hot Water Preheated in a 30-gallon glass-lined cylindrical tank within bin-of-stones. Final heating is by electrical heating elements in a separate 80-gallon stone-lined tank. In 1976-77 domestic hot water was 30 percent solar-heated.

Cooling in Summer None. The collector is not vented in summer, yet the stagnation temperature remains below 180°F because the panels, being vertical, receive relatively little radiation then.

Problems and Modifications Because the interior construction of the house is lightweight, the thermal capacity is small. On sunny days in winter the south portion of house (including clerestory) tends to become too hot. This difficulty was largely overcome by arranging to have warm air in the upper part of the clerestory circulated to cooler portions of the house. Initially, one motorized damper performed incorrectly. It was found to have been miswired at the factory. The wiring was soon corrected. During construction of the collector the importance of making the system airtight was kept in mind; careful tests have shown that the system is in fact airtight. Because the auxiliary heater is electric, determining the amount of auxiliary heat supplied is a simple matter.

Solar engineering: Total Environment Action, Inc. *Designer and builder:* CHI Housing, Inc. *Owner:* Anonymous. *Funding:* Private.

Solar heating system employs two large vertical air-type collectors and a 120-ton bin-of-stones

Norwich   44°N
(near Hanover, N.H.)

Building This U-shaped wood-frame house has two bedrooms, a partial basement, and no garage. The north wing and central region include kitchen, dining room, family room and bedrooms. The south wing is mainly a high-ceilinged living room. The lower parts of the outer walls consist of 8 inches of concrete insulated on the outside with 2 inches of sprayed-on urethane foam. The roof is insulated with 7 to 14 inches of fiberglass batts. The windows are covered at night by interior insulating shutters. The house faces exactly south. The "U" opens toward the east, defining a courtyard or terrace that is protected from the prevailing west wind.

> **Building:** 2-story, 2500 sq. ft.
> **Collector:** 860 sq. ft., air type
> **Storage:** 120-ton bin-of-stones
> **% Solar-heated:** 50 (predicted)

Collection The 860 square feet of vertical, air-type collector is distributed on two areas: the vertical south faces of the two wings. The south wing portion is 20 by 20 feet and is recessed 30 inches at top and sides; thus wind here is reduced and, in summer, much shading is provided. Glazing is double and consists of Kalwall Premium Sun-Lite, a polyester-reinforced fiberglass. Behind the glazing there is 1 inch of air and a .030-inch aluminum sheet with Alcoa selective black coating. This sheet is supported by 1-inch aluminum angle strips 6 inches apart. Blower-driven air travels at a speed of 500 feet per minute in the 1-inch space, defined by these strips, behind the black sheet. The back of the collector is separated from the insulated wall of the house proper by a 1-inch layer of fiberglass. The glazing support strips make a pattern of 2-by-2-foot squares. The collector on the other wing is generally similar but is specially shaped to minimize

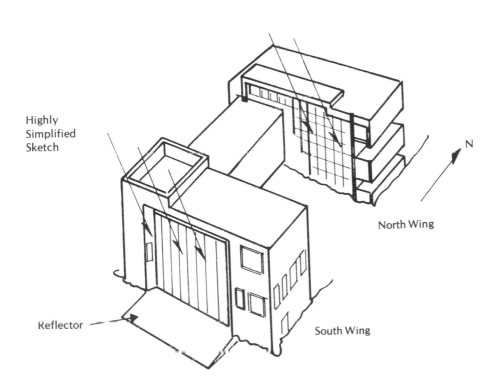

Highly
Simplified
Sketch

N

North Wing

Reflector

South Wing

shading in the afternoon. At the base of the south collector there is a 35-by-25-foot reflector consisting of a bed of white marble chips. Snow on the ground adds to the amounts of solar energy captured by the two collector areas. The maximum rate of airflow for the collector as a whole is 1050 cfm.

Storage   The 120 tons of 1-to-1½-inch diameter stones is contained in a rectangular bin 38 by 14 feet by 4½ feet high. The bin is situated under the central and northwest parts of the house and is largely below grade. The stones rest on steel screens supported by a labyrinth-and-channel system defined by concrete blocks resting on a sand bed. Collector air enters the top of the bin from both ends. An aluminum-foil vapor barrier and 7 inches of fiberglass are incorporated in the cover of the bin. The rooms are heated by forced flow of air that is circulated upward through the bin. The thick, externally insulated walls of the house contribute to the thermal capacity of the house as a whole.

Auxiliary Heat Source   Much use is made of an Austrian-type metal-and-ceramic wood-burning stove which may be fed as infrequently as every twelve hours and has very high efficiency. A fireplace is used also. Electrical resistance heaters situated in the duct leading to the bin-of-stones may be operated at off-peak hours only.

Domestic Hot Water   Preheated by bin-of-stones. Final heating is electric.

Cooling in Summer   None.

Problems and Modifications   During the first two years of operation, many small leaks in collectors and ducts were discovered. Because of the leaks, the solar heating system provided only about 30 percent of the winter's heat need. By the fall of 1977 nearly all of the leaks had been found and stopped, and the percent-solar-heated figure was expected to rise to 50. Late in 1977 the insulation of the storage system was being modified to reduce flow of heat to the (externally insulated) concrete foundation walls. It is believed that a considerable amount of heat was lost via those walls and via frame members near those walls.

Solar engineering: Total Environmental Action. *Architect, owner, and occupant:* C. S. White, Jr., of Banwell, White, and Arnold, Inc. *Mechanical engineer:*   J. H. Bates. *Performance studies:*   By A. O. Converse of Dartmouth College.

System employing combination of water-type collector and heat-pump

Building This three-bedroom wood-frame house has a basement and small attic, but no garage. The building is well insulated, mainly with sprayed urethane foam. The insulation on walls and ceilings is equivalent to 6 inches of fiberglass and 9 inches of fiberglass, respectively. The basement walls are insulated externally with the equivalent of 3 inches of fiberglass. There are no windows on the north side of the house. The windows on the other sides are triple-glazed. A small solarium, on the south side, has a total of 50 square feet of window and skylight area; at night insulating shutters are used.

Collection The collector is installed on two roof areas, each sloping 46 degrees. Eight of the sixteen collector panels are mounted on the

Quechee 43½°N (a 7500-degree-day location) (near Hanover, N.H.) Lot 7177, Quechee Lakes

Building: 3-story, 2300 sq. ft.
Collector: 400 sq. ft., water type
Storage: 2400 gals. water
% Solar-heated: See text

central roof and eight on the west roof. These are Grumman Sun-stream 60F panels, 110 by 40½ inches by 9 inches thick and 123 pounds in weight. The aluminum absorber plate consists of several side-by-side strips, or fins, of .05-inch aluminum. Each strip is about 5 inches wide and 9 feet long. The long edges of each strip are sharply angled and channeled so that it may be clipped firmly to the next strip and to a 3/8-inch-diameter tube of Type L copper that carries the coolant liquid. The tubes are 4¼ inches apart on centers and there are eight tubes in all. Because the tubes are clipped, rather than soldered, to the strips, they are free to slide relative to the strips, to relieve thermal expansion stresses. The coolant, a 50-50 solution of demineralized water and buffered ethylene glycol, flows along the tubes in parallel. The black coating used is nonselective. The glazing in single, consisting of a 1/8-inch-thick sheet of polymethyl metha-crylate. The sheet is slightly curved, cylindrically, to increase stiffness. The long edges fit into slots in the aluminum framing strips, and the slots provide room for thermal expansion. The panel backing includes 2 inches of fiberglass. The panels are hydraulically connected in tandem pairs lying end to end, with coolant flowing through the two panels in series. The eight pairs, however, are hydraulically in parallel. The coolant is circulated through the system at 10 gpm by a ⅓-hp centrifugal pump. Heat is delivered to the storage system via a heat exchanger.

Storage The 2400 gallons of water is kept in a rectangular poured-concrete tank with inside dimensions 16 by 6½ feet by 4 feet high. The tank is insulated on the outside with 2 inches of Zonite. The rooms are heated by forced air distributed by ducts. When the storage tank is hotter than 90°F the forced air system derives its heat directly from water from the tank. When the tank temperature is below 90°F, which is normally the case in midwinter, a York Triton DW-20H-B, 20,000 Btu/hr heat-pump is brought into play as intermediary. It derives its heat from the water in the tank and typically, in midwinter, cools that tank down to the 40°-to-80°F range. The typical COP of the heat-pump is 3. The heat-pump derives its heat from the storage tank only; no heat is derived from outdoor air or other source. The heat-pump is *not* used as auxiliary heat source. It is not used at all in summer.

Percent Heated by combination of solar heating system and heat-pump: 60. Percent heated by solar energy alone: 40.

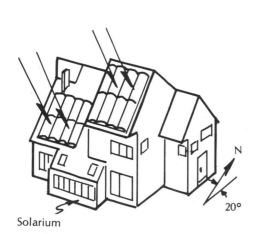

Solarium

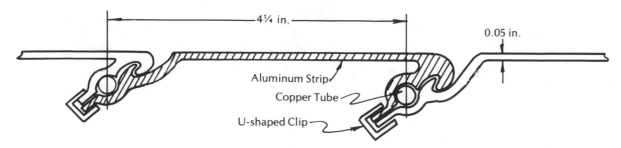

Cross section of a portion of the panel, showing several extruded aluminum strips joined together with U-shaped clips. The 3/8-inch-diameter copper tubes are clamped in the channels provided. The drawing is somewhat schematic and is not to scale.

Auxiliary Heat Source   An 85,000 Btu/hr oil furnace. Also a Heatilator fireplace at the west end of the living room.

Domestic Hot Water   In the coldest months the domestic hot water is heated solely by off-peak electrical power. In warmer months, when more solar energy is collected than is needed for space heating, the domestic hot water is preheated by the solar heating system by means of a wraparound, fail-safe heat exchanger applied to the pertinent 80-gallon tank. The collector then serves *only* to heat the domestic hot water. Such apportioning of the solar energy collected is dictated by the extremely low cost here of off-peak electrical power.

Cooling in Summer   None needed, none provided.

Note concerning low-temperature operating mode.   Many features of the design have been selected so as to insure good performance at low collector temperature, keep collector stagnation temperature low, and keep cost low. For example, the following design features would *not* be appropriate to *high*-temperature collector operation: single glazing, use of nonselective black coating, absence of insulation at exposed edges of panels, use of a panel-backing insulation only 2 inches thick. The reasons that low-temperature operation is effective are (1) when the collector operates at low temperature its collection efficiency is especially high and (2) the heat-pump is always available, when needed, to raise the temperature of the water sent to the heat distribution system. The cooperative use of solar system and heat-pump is a central feature of the overall design.

Problems and Modifications   The original collector panels, put into operation in the spring of 1976, were Sunstream 50A and employed an Olin Brass Co. Roll-Bond aluminum sheet. They were replaced in November 1977 with a superseding, superior design of panel called Sunstream 60F employing an extruded aluminum fin system. The original system worked well except for some air-binding that occurred as a result of (1) air in-leak at one imperfectly made connection of plastic CPVC tubing and (2) inadequate means for bleeding air from the system. Copper, rather than CPVC, tubing is now used, and no in-leakage of air has occurred.

Solar engineering: Grumman Energy Systems, Inc. (Kenneth Speiser, Edward Diamond, and others). *Architect:* Blue Sun Ltd. *Builder and owner:* Terrosi Construction Company.

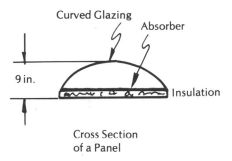

Cross Section
of a Panel

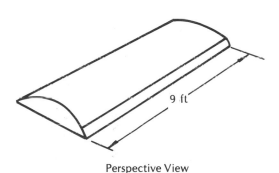

Perspective View
of a Panel

Ferrum   37½°N
(a 4200-degree-day
site)
(35 mi S of Roanoke)
Two miles north of
Ferrum College

Building: 2-story, 1350 sq. ft.
Collector: ⎫
                 ⎬ Passive system
Storage: ⎭
% Solar-heated: 50-85 (predicted)

Passively heated house employing several 32-inch-long, two-story-high masonry walls angled to admit solar energy in the morning but intercept and store it in the afternoon

Building   This two-story, three-bedroom house, 32 feet long and 24 feet wide, has no basement, attic, or garage. The external walls, and also the internal load-bearing walls, are of concrete blocks or cinder blocks tamped full of sand. Earth berms protect the north and east sides of the first story. The floor of the first story is a 4-inch concrete slab resting on 4 inches of gravel. The outer walls are insulated with 3 inches of urea-formaldehyde foam between the concrete-block wall and the external siding. The roof is insulated with 7½ inches of urea formaldehyde foam. The foundation walls are externally insulated with polystyrene foam. The south window area is 350 square feet. The combined area of the windows on the east and west sides is 70 square feet. The window area on the north side is negligible. In winter, all of the windows are double-glazed. Where view is important, glass is used; elsewhere Kalwall Sun-Lite is used. The first story includes three bedrooms and a bath and the second story includes kitchen-dining-living room, library, and bath; this allocation of the space has the desirable consequence that the sleeping area tends always to be cooler than the kitchen-dining-living area.

Passive Solar Heating System   Solar radiation enters the building via 350 square feet of south windows, most of which constitute the vertical, two-story-high south window-wall. The clerestory windows, which admit radiation to the north half of the second story, have an

area of 50 square feet. Storage is provided by walls of sand-filled concrete blocks, a concrete floor slab, and other massive components. The total mass is 100 tons. Of special importance is a set of five vertical masonry walls, each two stories high, situated within the south rooms and close to the south windows. Each of these short walls is 15 feet high, 32 inches long, and 12 inches thick. Each is at 45 degrees to the south face of the house and lies in a southeast—northwest plane. The short walls are 4 feet apart on centers. They serve as a set of optical louvers and as a storage system. Specifically, they (1) admit much light to the rooms, (2) permit occupants to obtain a view to the southeast, south, and southwest (but not the west-southwest), (3) allow much morning solar radiation to penetrate deep into the rooms, warming them, (4) intercept and absorb most of the afternoon radiation, thus reducing the tendency for the rooms to become too hot, (5) store much heat in the afternoon, (6) release much heat at night, thus keeping the rooms warm then, and (7) serve as supports for vertical curtains or shutters that, at the end of the day, may be slid into position to insulate the large south windows. Note that the set of short walls serves the main function of a Trombe wall but avoids many of the drawbacks of such a wall.

Auxiliary Heat Source A small, high-efficiency wood-burning stove on the second floor.

Domestic Hot Water Not solar-heated.

Cooling in Summer There is no formal cooling system. The 40-inch eaves of the south roof exclude solar radiation from the upper south windows, and the lower south windows are shaded by an exterior shade that is stored, when not in use, close below eaves. Much natural ventilation, assisted by the clerestory windows, is provided. The massive walls and massive first-story floor help keep the temperature rise small.

Designer, builder, owner, occupant: James Bier. *Cost, not including land or the labor provided by the owner:* $10,000. *Funding:* Private.

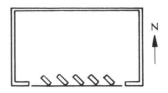

Plan view showing the 45° orientation of the five short masonry walls

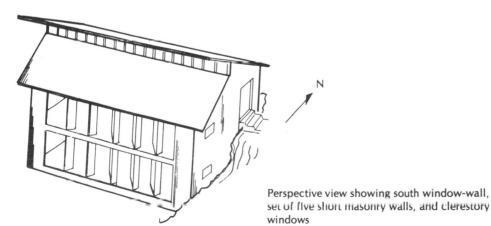

Perspective view showing south window-wall, set of five short masonry walls, and clerestory windows

Madeira School Science Building

Greenway   39°N
(a 4000-degree-day
site)
(20 mi. NW of
Washington, D.C.)

Building: 1½-story, 9000 sq. ft.
Collector: 4000 sq. ft., with oil coolant
Storage: 10,000 gals. water
% Solar-heated: No value available

9000-sq.-ft. science building on north slope of hill is solar-heated by a 4000-sq.-ft. collector employing, as coolant, a special oil

Building   This modern-style science building, on the north slope of a hill, includes classroom-laboratories for biology, chemistry, and physics; also a greenhouse, space for animals, and a basement photographic laboratory. The upper (main) story is trapezoidal; the north and south walls are 201 feet and 70 feet long respectively; the building width is 47 feet; the floor area is 6000 square feet. Most of the windows are on the north side. Adjacent to the south side of the building there is a walkway that extends beneath the overhang of roof and collector. Sliding partitions between main rooms permit flexible use. Laboratory equipment is stored in portable cabinets hung on racks along the north wall and transported by dollies. There is an abundance of sink stations, water outlets, electrical outlets, and safety equipment. The lower story is half crawl space and half useful space, the latter consisting of 3000 square feet of mechanical rooms, printing shop, and photographic laboratory. There are a few windows on the north side, and the south side is below grade.

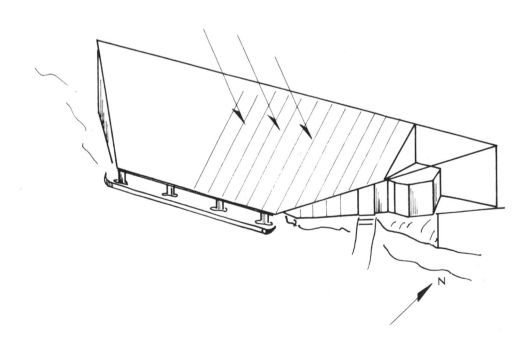

Collection The collector covers almost all of the trapezoidal roof, which slopes 26 degrees. The roof is of wood waterproofed with a silicone-base membrane. The collector consists of Olin Brass Company aluminum Roll-Bond sheets, each 104 by 27 inches, with integral passages for the coolant liquid, which is Exxon heat-transfer oil Caloria HT-43. The black coating on the aluminum sheets is nonselective. The panels are single-glazed. A centrifugal pump forces the coolant to flow upward within the collector panels, at a rate of 72 gpm, to a header pipe on the ridge of the roof. Heat is imparted to the storage system by a heat exchanger.

Storage The 10,000 gallons of water is contained in an insulated, horizontal, cylindrical, 8-foot-diameter steel tank situated in the mechanical room. The tank temperature is 130°F typically, the maximum value being 180°F. Heat is delivered to the rooms by hot-water circulation.

Auxiliary Heat Source Steam boiler in adjacent building.

Cooling in Summer Not yet planned. Suitable space is available.

Swimming Pool Heating The collector is used to heat the swimming pool, closely adjacent to the south, in spring and fall when the building heating requirement is small.

Problems and Modifications Balancing the control system took much time.

Solar engineering and general engineering: Flack and Kurtz, Inc. *Architect:* Arthur Cotton Moore Associates (esp. Kent Abraham). *Builder:* Commercial Industrial Construction, Inc. *Funding:* Private.

House 100% heated by the combination of a two-collector solar heating system and a heat-pump

Building The three-bedroom wood-frame house consists of three connected parts, or modules: south, with three bedrooms; north, with kitchen, living room, and dining room; and east, with the attached one-car garage. The air-lock main entrance is in a hallway between the north and south modules. There is no basement. The insulation used consists mainly of urea tripolymer foam; the walls are insulated to R-26 and the ceilings or roofs to R-36. The windows, with a total area of 300 square feet, are double-glazed and are covered at night by retractable external shutters. The house includes many special devices for saving heat and water.

Collection The collector is on two 58-degree-sloping roofs. In all, there are eighteen Chamberlain Manufacturing Company panels, each 8 by 3 feet. (Two of these are for the domestic hot water system.) The heart of a panel is a pair of steel sheets that are formed and welded together to produce a waffle-type pattern, with coolant flowing between the two sheets. The black chrome coating is selective. The glazing is single and consists of a 1/8-inch sheet of low-iron glass. The panel backing includes 3 inches of fiberglass. The water-and-antifreeze coolant is circulated at 6 to 9 gpm.

Hampton   37°N
(a 4500-degree-day location)
(in SE corner of state)
At Langley Research Center

Building: 1-story, 1500 sq. ft.
Collector: 370 sq. ft., water type
Storage: 1900 gals. water
% Solar-heated: See text

Storage The 1900-gallon tank, a rectangular concrete septic tank that has been waterproofed, is insulated externally with 2 inches of foam. The rooms are heated by a multizone forced-air system. The coil of the forced-air system receives hot water directly from the collector if this water is hotter than 95°F; if it is cooler than this, heat is supplied by a 1-hp Florida water-to-air heat-pump, which extracts heat from the storage tank water if this is hotter than 60°F, otherwise from well water.

Percent Heated by combination of solar heating system and heat-pump: 100. The percent attributable to the solar heating system itself is 80.

Auxiliary Heat Source A high-performance wood-burning fireplace that supplies hot air to the rooms and, by means of a coil-grate, supplies hot water to the storage tank.

Domestic Hot Water This is preheated by two of the Chamberlain collector panels mentioned above. Final heating is electric. A special tank for the solar-preheated water is provided.

Cooling in Summer The heat-pump, operated in reverse mode, cools the house; the rejected heat is transferred to well water or is dissipated at night by radiation to the sky. The amount of cooling needed is small because of the 4-foot eaves above the south windows, the excellent insulation of the building, and the large louvered ventilation areas serving the spaces beneath the roofs.

Research study: Moore Grover Harper (esp. W. H. Grover) and Forrest Coile and Associates (esp. H. R. Cuppett). *NASA solar engineers:* Robert Basford and colleagues. *General engineering and architecture:* Forrest Coile and Associates (esp. H. R. Cuppett and D. DiBlazio) assisted by Moore Grover Harper (esp. W. H. Grover). *General planning, owner, user, and source of funds:* NASA Langley Research Center.

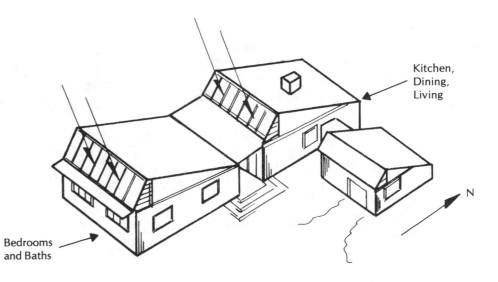

80% solar-heated by 950-sq.-ft. trickling-water-type collector and storage system employing 50 tons of stones and 1600 gallons of water

Luray   38½ °N
(75 mi WSW of
Washington, D.C.)
In Egypt Bend Estates

Building This two-bedroom wood-frame house, 40 by 26 feet, has a storage attic and a 750-square-foot basement. There is a carport at the west end of the house. The walls and ceilings are insulated to R-11 and R-13 respectively. There are no windows on the south side because the collector preempts that side. The window areas on the other sides are moderate. Anderson double-glazed windows are used. The house faces exactly south.

Collection The trickling-water-type collector, based largely on the schemes used by H. E. Thomason, is on the south roof and slopes 45 degrees. The collector is 44 feet long and 22 feet wide. The glazing is single and consists of double-strength glass panes 24 by 22 inches. The absorber is a sheet of corrugated aluminum with a nonselective black coating. Water is distributed to the valleys of the corrugated sheet by a pipe that runs along the ridge of the roof; it is collected at the lower edge of the roof by a gutter. A flowrate of 20 gpm is maintained by a ½-hp centrifugal pump. No antifreeze is used; the system drains automatically when the pump stops. The hot water from the collector flows directly into the storage tank. There is no heat exchanger.

Storage The 50 tons of 3-to-6-inch-diameter stones is contained in a rectangular concrete-block bin in the basement. The bin is 24 by 9 feet and 7½ feet high. Within the bin, and surrounded by stones, is a horizontal, cylindrical 1600-gallon steel tank 19 feet long and 4 feet in diameter. Water is circulated directly from this tank to the collec-

> **Building:** 1-story, 1040 sq. ft.
> **Collector:** 950 sq. ft., trickling-water type
> **Storage:** 50 tons stones and 1600 gals. water
> **% Solar-heated:** 80 (predicted)

tor. Heat from the tank warms the stones. When the rooms need heat, a ½-hp blower circulates room air through the bin-of-stones.

Auxiliary Heat Source   A fireplace.

Domestic Hot Water   This is not heated by the solar heating system.

Cooling in Summer   None.

Problems and Modifications   The wooden 2x4 strips that are situated just above the black absorber sheets and support the glazing have deteriorated somewhat. The builder believes that aluminum framing would have been more durable.

Architect, builder, owner: J. Wesley Poland, Contractor, Inc. *Architectural assistance:* Provided by M. Cullers and E. C. Lynch. *Cost of house and solar heating system:* About $32,000.

Porch
(above)
Carport
(below)

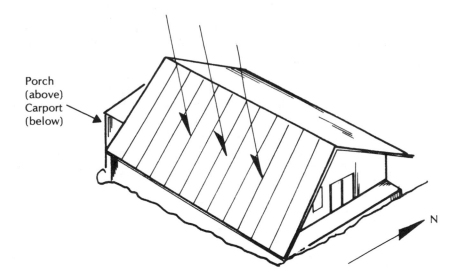

Bunn House

565-sq.-ft. air-type collector and 22-ton bin-of-stones provide 55% solar heating

Madison   43°N
1906 Capital Avenue,
on Lake Mendota

Building   This is a three-bedroom, multilevel, 2½-story wood-frame house with attic, basement, and built-in one-car garage. The walls are insulated with 3½ inches of fiberglass and the attic is insulated with 10 to 12 inches of fiberglass. Total area of windows is 440 square feet; most of the windows are triple-glazed.

Collection   The air-type collector, on a roof sloping 58 degrees, includes twenty-nine Solaron panels, each 6½ by 3 feet. The absorber consists of a 20-ounce sheet of steel with integral fins and 5/8-inch channels on the underside. The black coating is nonselective. The panels are double-glazed with 1/8-inch sheets of Fourco low-iron tempered glass and are hermetically sealed. The fiberglass backing is 3½ inches thick. Air is circulated upward in the channels at 1130 cfm by a ¾-hp blower situated in the subbasement. The supply and return ducts have a cross section 18 by 16 inches.

Storage   The stones, of 1½-inch diameter, are contained in a rectangular poured-concrete bin, 22 by 5 feet by 5 feet high, situated in the subbasement. The bin is insulated with 2 inches of Styrofoam. The rooms can receive hot air directly from the collector or from the bin-of-stones; the air is circulated by the blower mentioned above. Dampers control the flow. Hot air from the collector circulated through the bin travels downward, and when room air is circulated, it travels upward through the bin.

Building: 2½-stories, 1900 sq. ft.
Collector: 565 sq. ft., water type
Storage: 22-ton bin-of-stones
% Solar-heated: 55

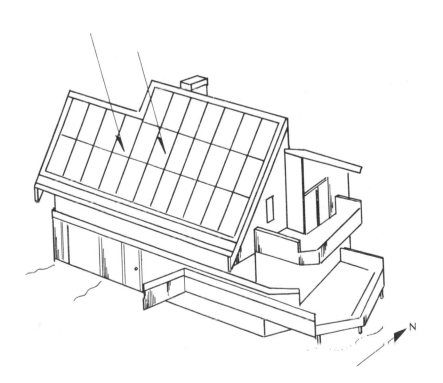

Auxiliary Heat Source A gas furnace rated at 110,000 Btu/hr and a forced-air distribution system. The auxiliary system has its own ¾-hp blower.   There are also two wood-burning Jøtul stoves.

Domestic Hot Water   This is preheated by the solar heating system.

Cooling in Summer   Cool night air is circulated through the bin and, during the day, room air is circulated through the bin.

Solar engineering: Solaron Corporation. *Architect:* C. O. Matcham, Jr. *Heat-loss calculations:* S. A. Klein of the University of Wisconsin Solar Energy Laboratory. *Builder:* M. Uselman. *Owner and occupant:* George Bunn. *Funding:* Private.

Very simple, very low-cost, passively solar-heated house with large integral greenhouse

Building   This 64-by-15-foot, owner-built house has no attic, basement, or garage. Pole construction was used. The flat monoslope roof slopes 10 degrees downward toward the north. Walls and roof are insulated with 6 inches of fiberglass. The south wall, serving the greenhouse, is described below. The windows on the other walls are of negligible area and are double-glazed. Most of the house has a 3½-inch concrete floor slab which rests on a 2-inch layer of beadboard. The edges of the slab are insulated with 4 inches of beadboard. The building faces exactly south.

Passive Solar Heating System   Solar radiation enters through the windows of the greenhouse, which is 64 feet long and 5 feet wide. The row of windows is 64 feet by 6 feet high; the area is about 350 square feet. The windows slope 60 degrees and are double-glazed with .004-inch polyethylene sheets that cost 26 cents per square foot. The sheets are ¼ inch apart and are held by wooden frame members and battens. The sheets are replaced each year. The greenhouse growing area is divided into two long slender strips, one on grade and one above grade. The on-grade strip, 64 x 3 feet, is immediately adjacent to the greenhouse window. The earth here is 16 inches deep and

Building:	1-story, 1000 sq. ft.
Collector:	⎰ Passive system; see text
Storage:	⎱
% Solar-heated:	60

rests on 2 inches of beadboard; the sides are insulated with 2 inches of beadboard. The above-grade strip, immediately adjacent to the north, is 16 inches higher and is 64 by 2 feet. Its south edge is of 4-inch-thick concrete blocks. Immediately north of this south edge there is a row of seven 40-gallon water tanks (salvaged domestic hot water heater tanks) partially exposed to the incident solar radiation and partially buried in the earth. Each tank is horizontal, with its axis running east-west. The tanks contain, together, 280 gallons. At night the greenhouse windows are insulated with rectangular, 2-inch-thick panels of beadboard affixed to the indoor sides of the windows. Solar energy is absorbed and stored directly in the greenhouse earth, the concrete blocks, and the water-filled tanks, and much heat is absorbed indirectly by the concrete floor slab. If the house becomes too warm, hot air is vented to the outdoors. Moist earth keeps the house humidity comfortably high. The large growing area permits growing large quantities of vegetables.

Auxiliary Heat Source   A sheet-metal wood-burning stove.

Domestic Hot Water   This is not heated by the solar heating system.

Cooling in Summer   Natural ventilation is used. To increase the ventilation, some of the greenhouse windows are removed and replaced by screens. On unusually cold nights in summer the screens are covered.

Designers and builders: David Kruschke, P. Urban, and K. Funk. *Owner and occupant:*   David Kruschke. *Cost of house and solar heating system:* $10,000.

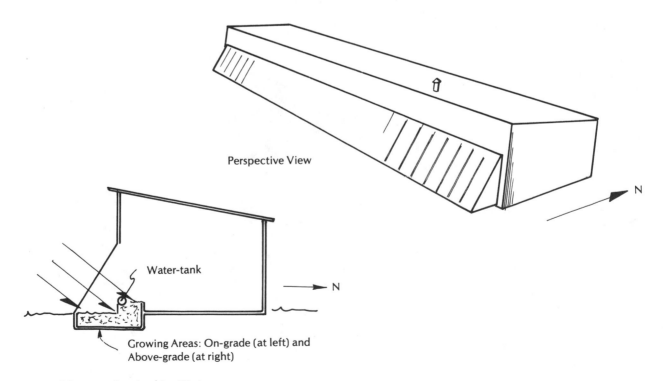

Perspective View

Water-tank

N

Growing Areas: On-grade (at left) and
Above-grade (at right)

Vertical Cross-section, Looking West.

Cheyenne House

3800-sq.-ft. house 45% solar heated by 470-sq.-ft. air-type collector and 15-ton bin-of-stones

Cheyenne   41°N

Building   This four-bedroom, wood-frame house has a full base-ment, a small attic space, and an attached two-car garage. The house is recessed into a south-sloping hillside. Walls are insulated with 5½ inches of fiberglass, and the ceiling or roof is insulated with 9½ inches of fiberglass. The interior face of the basement wall is insu-lated with 2 inches of Styrofoam. The window area is modest, and all windows are double-glazed. The south windows are deeply recessed. In the center of the house there is a 150-square-foot atrium, open to the sky. Many of the rooms have windows opening to the atrium. The house faces exactly south.

Building: 2-story, 3800 sq. ft.
Collector: 470 sq. ft., air type
Storage: 15 tons of stones
% Solar-heated: 45 (predicted)

Collection   The collector, which is mounted on the south roof and slopes 53 degrees, includes twenty-four Solaron Corporation panels, arranged in two rows. A ¾-hp blower in the air-handling device in the basement collects air from the uppermost part of the second-story rooms (ordinarily the hottest air in the house), sends this air to the lower manifold of the collector, and thence drives it upward through the collector at 950 cfm. The collector is double-glazed with 1/8-inch tempered glass. Because the south windows are deeply recessed, they collect only a moderate amount of radiation.

Storage   The 15 tons of 1½-inch-diameter stones is housed in a rectangular, 8-by-8 foot bin, 6 feet high, situated in the basement. The bin is insulated with 6 inches of fiberglass. When the rooms need heat, air from the lowest rooms is driven upward through the bin by the above-mentioned blower.

Auxiliary Heat Source   Natural-gas-burning furnace with forced-air distribution system.

Domestic Hot Water   Preheated by solar heating system by means of an air-to-water heat exchanger in the duct carrying hot air from the collector.

Cooling in Summer   None, other than by ventilation assisted by a wind-turned turbine on the ventilation shaft. The need for cooling is small, inasmuch as the house is at high altitude and is well insulated, and the south windows are deeply recessed.

Solar designer and architect: Crowther Solar Group (especially R. L. Crowther; also P. Karius). *Builder:*   Edeen Construction Company.

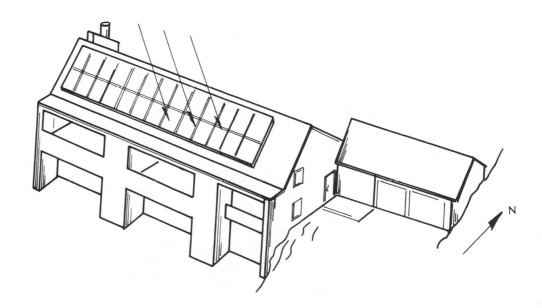

Solar-heated by 970-sq.-ft. water type collector

Coquitlam   49° N
(suburb of
Vancouver)

Building   The three-bedroom house has a full basement and a small attic. The walls, insulated with 3½ inches of fiberglass and 1½ inches of exterior Styrofoam, have an R-value of 16. The ceiling or roof, insulated with 8 inches of fiberglass, has an R-value of 27. The windows, of small total area, are triple-glazed.

Collection   The 54-by-18-foot collector is on the 37½-degree mono-slope south roof. Each of the twelve panels, made by Solarsystems Industries, Ltd., is 18 by 4 feet. The absorber, a sheet of 8-ounce copper, has a nonselective black coating of Nextel. The ¼-inch-ID copper tubes attached to the sheet are 5 inches apart on centers. The panels are double-glazed, the outer glazing consisting of a plastic sheet of fiberglass and polyester with a Tedlar coating, and the inner sheet, one inch from it, of Tedlar. The outer sheet is corrugated, the corrugations running up and down the roof. The coolant, which is water with no antifreeze, is circulated at 20 gpm by means of a ¼-hp centrifugal pump and drains automatically when the pump stops.

Storage   The 3000-gallon concrete tank, with 6 inches of insulation, is in the basement. The rooms are heated by a forced-air system.

Building: 2-story, 1500 sq. ft.
Collector: 970 sq. ft., water type
Storage: 3000 gals. water
% solar heated: 30 — 50 (predicted)

Auxiliary Heat Source   A 2½-ton General Electric air-to-air heat-pump is used. It extracts heat from outdoor air and delivers the heat to the forced-air system.

Domestic Hot Water   This is preheated by being circulated through a ¾-inch-diameter, 60-foot-long copper pipe coil inside main tank.

Cooling in Summer   The heat-pump is used in cooling mode.

Solar engineering and architectural design: Phillips Barratt, Engineers and Architects. *Solar panels supplier:*   Solarsystems Industries, Ltd., (E. W. Hoffman and others). *Builder:*   Belcar Industries, Ltd. *Assisting agencies:* B. C. Hydro; Housing and Urban Development Association of Canada. *Funding and original owner:*   Pacific National Exhibition.

Trousdell House

90 % solar heated by combination of trickling-water-type collector on workshop roof and large area of vertical windows

Duncan   49° N
(36 mi N of Victoria)
3878 Old Lake
Cowichan Road

Building   The three-bedroom house has a small attic and a crawl space, but no basement. The walls are insulated with 6 inches of fiberglass to R-20, and the ceiling or roof is insulated with 10 inches to R-30. The south window area is 160 square feet, and the total window area on the other three sides is 100 square feet. All of the windows are double-glazed, and most of them are covered at night by 2-inch-thick indoor shutters made of rigid foam; at the start of the day the shutters are folded out of the way. The north side of the house is protected by an earth berm. Just to the east of the house there is an unheated workshop, the roof of which supports the collector.

> **Building:** 1-story, 1500 sq. ft.
> **Collector:** 800 sq. ft. trickling-water type
> **Storage:** 5000 gals. water
> **% Solar-heated:** About 90

Collection   The 40-by-20-foot trickling-water-type collector, mounted on the workshop roof, slopes 67 degrees. The heart of the collector is a corrugated galvanized steel sheet, with the corrugations running up and down. A 1-inch-diameter distribution pipe that runs along the ridge of the roof supplies water, via a series of 160 holes, each 1/8 inch in diameter, to each valley of the corrugated sheet. On being collected by a gutter at the bottom, the water flows to the storage system. No heat exchanger is used. There is no antifreeze; the water drains automatically whenever the 20 gpm, 1/3-hp centrifugal pump stops. The outer glazing layer is corrugated Filon, a polyester-filled fiberglass, and the inner layer is of .004-inch Tedlar, with a 1/4-inch airspace between them. The airspace between the inner layer and the galvanized sheet is also 1/4 inch. The nonselective black coating on the corrugated sheet consists of Isoclad protected by Palidux waterproofing agent. The control system includes a differential thermometer.

Storage   The steel-reinforced, poured-concrete tank, situated below grade under the house, has the inside dimensions 10 by 10 feet by 9 feet high. It is waterproofed on the inside and is insulated on the outside with Styrofoam. Heat is distributed to the rooms by a fan-coil system. A moderately large amount of heat is contributed passively by the 160 square feet of vertical south windows. The active system contributes far more than the passive system to the total solar heating which provides about 90 percent of the winter heat need.

Auxiliary Heat Source   A fireplace. There is also a 15-kw electric immersion heater in the main tank, but it has not been used.

Domestic Hot Water   Preheated in a 30-gallon copper tank within the main tank.

Cooling in Summer   None. Solar radiation is excluded by the large eaves.

Problems and Modifications   One temperature sensor failed when first installed and was replaced. Some water leaked through the absorber-plate system and onto the wooden sheathing beneath it; the leakage occurred at locations where nails had been driven through the absorber plate to secure it to the roof. Washers and neoprene gaskets were provided but they failed to prevent the leaks. The difficulty was overcome by temporarily removing the glazing and applying a silicone sealant. Plans have been made for installing a 22-by-14-foot greenhouse that would enclose the south-side bay-window area of the house proper; eight water-filled 55-gallon steel drums would provide storage.

Solar designers: C. Mattock (architect) and P. Summerlin (engineer). *Owner and resident:*   T. Trousdell. *Monitoring:*   B. C. Hydro and Power Authority. *Cost of solar heating system:*   Said to be about $6600, that is, about $5000 more than the cost of a conventional forced-air system.

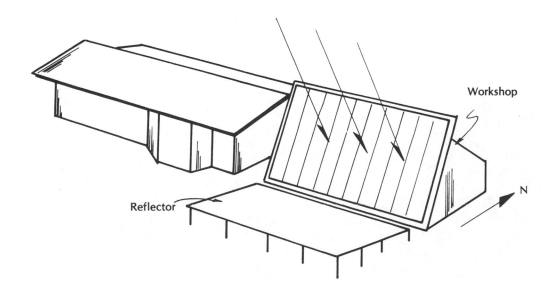

Retrofitted solar heating system provides 50% of winter heat need

Surrey 49° N
(near Vancouver)
5511 128 St.

Building The house was built in 1968 and the solar heating system was installed in 1971. The 1400-square-foot first story is fully heated and the 1300-sq.-ft. basement is partially heated.

Collection The collector is on the roof that slopes 58 degrees. The .005-inch copper sheet has a non-selective black coating. Soldered to this sheet are ¼-inch-diameter copper tubes spaced 6 inches apart. Double-glazing is used; the outer sheet, ¾ inch from the inner sheet, is of double-strength glass, and the inner sheet, ¾ inch from the black copper sheet, is of single-strength glass. The collector backing includes 6 inches of fiberglass. The water used contains no antifreeze; it drains automatically when the pump stops.

Storage Two vertical, cylindrical tanks, of 500-gallon and 300-gallon capacity, are used. They are uninsulated but are situated in a small, insulated room in the basement. Gravity-convective circulation of warm air from this small room keeps the main-story rooms warm. Heat is supplied to a swimming pool via a heat exchanger in one of the two storage tanks.

Auxiliary Heat Source Electric.

Domestic Hot Water This is preheated by the main solar heating system.

Building: 1-story, 1400 sq. ft.
Collector: 460 sq. ft., water type
Storage: 800 gals. water
% Solar-heated: 50

Cooling in Summer   None.

Problems and Modifications   In the first years of operation these troubles arose: breakage of some glass covers; freezing of the water in the pipes as a result of imperfect draining; failure of the PVC pipe carrying water from the collector to the storage system on a sunny day when the flow of water stopped and steam developed. The troubles were soon overcome and in recent years the performance has been uneventful.

Solar engineer and owner: E. W. Hoffmann.

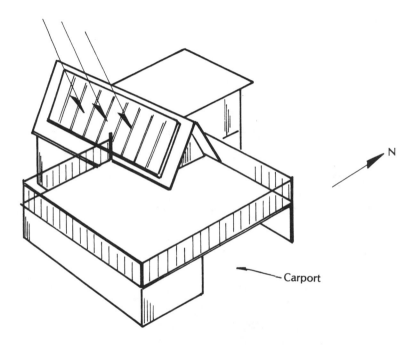

60% solar heated by water-type solar heating system

Mississauga 44° N
(suburb of Toronto)
(a 4100-degree-day site)
2940 Quetta Mews, Meadowvale

Building The three-bedroom brick house has a full basement but very little attic space. At the northwest corner of the house there is an attached one-car garage. The plan dimensions of the house are 34 by 27 feet. The house faces exactly south. Of the 265-square-foot window area, 70 percent faces south and 11 percent north. All of the windows are double-glazed. The stud walls are insulated with 6-inch fiberglass batts and the masonry walls include 3 inches of rigid insulation. The ceilings include 8-inch fiberglass batts. The main vestibule is of air-lock type.

> **Building:** 2-story, 1450 sq. ft.
> **Collector:** 690 sq. ft. water type
> **Storage:** 4000 gals. water
> **% Solar-heated:** 60% (predicted)

Collector The collector, mounted on a roof that slopes 60 degrees, includes 33 Sunworks panels. Each is 7 by 3 feet and employs a copper sheet, with selective black coating, and copper tubes. The glazing is single. No antifreeze or inhibitor is used; the water, circulated at 17 gpm, is drained before freeze-up can occur.

Storage Two steel-reinforced, poured-concrete tanks, each of 2000-gallon capacity, are used and are situated in a corner of the basement. Each tank is 10 by 6 feet by 8 feet high; it has 4 inches of rigid insulation on top and sides and none on the bottom. Hot water from the collector can be sent to either tank or to both in series. Normally one tank is appreciably hotter than the other, and the hotter

tank supplies water for heating the rooms by a forced-air system. The cooler tank supplies water to the collector. When both tanks are colder than 120°F, a water-to-air heat-pump extracts energy from the two tanks and delivers the energy to ducts serving the rooms. The heat-pump is situated above one tank. (Note: the heat-pump does not transfer energy from one tank to the other, nor does it extract heat from outdoor air.) Although the house is about 60 percent solar-heated, a much larger percentage of the winter's heat need is provided by the combination of solar system and heat-pump.

Auxiliary Heat Source   Electric resistance heaters in ducts serving the rooms.

Domestic Hot Water   This is heated by the solar heating system at times of the year when this system has excess capacity.

Cooling in Summer   In summer the heat-pump operation is reversed: heat is taken from the room air and delivered to a stream of water that is to be discarded.

Problems and Modifications   Some moisture condensation in the collector panels occurred. The heat loss from the storage tanks was greater than expected, presumably because the bottoms of the tanks were not insulated and there is some groundwater nearby. Late in 1977 the electric heater system was revised so as to operate at off-peak times only.

Designer: Douglas Lorriman, assisted by D. H. Lee and B. Fergusson. *Architect:* Lee, Elken, Becksted, Paulsen, Fair. *Mechanical consultant:* Mechanical Consultants Wester Ltd. *Owner:* Mississauga Solar Demonstration Project Ltd. *Cost of solar heating system:* About $45,000 (estimated). *Cost of house as a whole,* including the solar heating system: About $145,000 (estimated). *Funding:* $60,000 provided by Ministry of Urban Affairs in connection with Canadian Urban Demonstration Project, for the solar heating system and trial operation. The balance of funding was private, with some help from corporations.

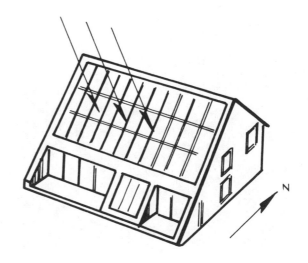

Stanley-Jones House

Very well-insulated house near-100% solar heated by 230-sq.-ft. water type collector

Osgoode   45° N
(20 mi S of Ottawa)
NW¼, Lot 27,
Concession 2

Building   The three-bedroom, wood-frame house has a full base-ment. Heated area, including basement, is 1850 square feet. First-story walls, which are externally faced with 6 inches of fieldstone, include 3½ inches of rockwool, 2 inches of Styrofoam, and other components; the overall R-value is 27. Ceilings are insulated to R-52. The basement walls, of concrete, are insulated on the outside with 3 inches of polystyrene foam, or, near the base of the walls, 1 inch. A 3/8-inch drainage space is left between concrete and insulation. All windows are double-glazed. There are two dormers on the south side of roof and one on the north side. The house faces exactly south.

> **Building:** 1½-story, 1850 sq. ft.
> **Collection:** 230 sq. ft., water type
> **Storage:** 6000 gals. water
> **% Solar-heated:** Near 100 (predicted)

Collection   The collector occupies the south roof space above the two dormers, and slopes 45 degrees. It consists of a single row of eleven Sunworks, Inc., panels, each 7 by 3 feet. Each has a copper absorber sheet that has a selective black coating, with a = .95 and e = .1. Glazing consists of a single sheet of high-transmittance 3/16-inch tempered glass. The backing includes 2½ inches of fiber-

glass. The coolant, water containing 50 percent ethylene glycol, is circulated at 4 gpm by a ¼-hp centrifugal pump. Heat is delivered to the storage system via a heat exchanger or is delivered directly to the rooms.

Storage   The horizontal, cylindrical steel tank, 16 feet long and 8 feet in diameter, is buried underground just to the west of the house. The tank is insulated with polystyrene foam: 12 inches on the top of the tank and 6 inches on the other surfaces. When the rooms need heat, water from the top of the tank is circulated to baseboard radiators in the rooms.

Auxiliary Heat Source   An oil-fired water heater and also a special fireplace that has many heat-saving features. The fireplace alone is capable of supplying 100 percent of the heat needed.

Domestic Hot Water   This is not heated by the solar heating system.

Cooling in summer   None needed, none provided.

Solar engineer, architect, builder, owner, occupant: M. L. Stanley-Jones.

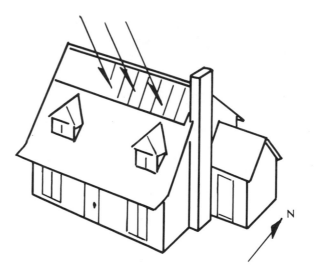

Outdoors

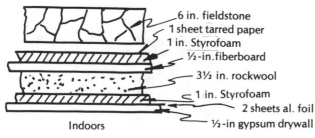

6 in. fieldstone
1 sheet tarred paper
1 in. Styrofoam
½-in. fiberboard
3½ in. rockwool
1 in. Styrofoam
2 sheets al. foil
½-in gypsum drywall

Indoors

Cross section of wall of first story

274

Large, multi-use building [dwelling, greenhouse, aquaculture space, etc.] solar-heated by a combination of several collection systems and storage systems

Building   The wood-frame building faces exactly south. Maximum length and width are 110 feet and 46 feet respectively. The building includes a dwelling, a general work area, and a main greenhouse.

The dwelling, at the west end of building, is a two-story structure with a small greenhouse, basement, attic, and two decks. The two main stories have a combined area of 1800 square feet. The lower story includes the kitchen-dining-living area. The upper story has three bedrooms. The small (220-square-feet) greenhouse opens onto the first-story living area. The basement includes three large, rectangular, water-filled thermal storage tanks and also a mechanical (utility) room. The attic contains a 120-gallon cylindrical tank for domestic hot water. There are two decks: upper and lower; they are at the west end of the dwelling.

The general work area, in the central north part of building, includes a lower central room used as workshop, a room just above this used both as a laboratory and as a center for monitoring equipment, a northeast room serving as barn and garage, and a loft just above this used as an office.

The main greenhouse, in the south central and southeast part of building, consists of a single long room, 80 by 24 feet, with its long axis east-west. The area is 2000 square feet. Along the central and southern parts of the room, plants are grown. Along its northern part

Little Pond   46°N
(an 8400-degree-day location)
(60 mi east of Charlottestown)
Souris Rural Route 4, Spry Point

Building: 1-to-2 stories, 7000 sq. ft.
Collector: Several systems; see text
Storage: Several systems; see text
% Solar-heated: 60 to 100; see text

there are two rows of vertical, cylindrical water-filled tanks, each 4 feet in diameter and 5 feet high and made of Kalwall Sun-Lite (translucent fiberglass-reinforced polyester), which contain fish. There are thirty such tanks in all. The tanks are not insulated; accordingly they accept or give out heat readily and thus help reduce changes in room temperature.

The building is well insulated. Typical wall insulation includes 5½ inches of fiberglass and (on outer side) 1 inch of extruded Styrofoam. Typical roof insulation includes fiberglass batts 12 inches thick. Foundation walls are insulated on their outer sides with 2 inches of Styrofoam. View windows are of modest size and are double-glazed. There are no windows on the north side of the building. The main entrance, on the north side of building, has an air-lock entryway.

The two toilets are of Clivus Multrum type. Two hundred yards to the north there is a windmill that can generate 7.5 kw of electricity.

Solar Heating System   The solar heating system includes a passive solar heating system for the small greenhouse and the main greenhouse, an active solar heating system for the dwelling, and a temperature-stabilizing system for the main greenhouse.

Passive solar heating system for small greenhouse and main greenhouse.   Both greenhouses are heated by solar radiation that passes through a large south-facing window array that slopes 40 degrees. The dimensions of the array are 102 by 18 feet and the area is 1800 square feet. The glazing is double; it consists of a Rohaglas cellular acrylic structure (U.S. name: Acrylite SDP) made by Chemacryl Plastics Ltd. The structure includes two translucent acrylic sheets, ½ inch apart, joined by hermetically attached acrylic partition strips ½-inch apart on centers. The 18-by-4-foot panels of this material have a transmittance of about 83 percent and a thermal resistance of R-1.7. The panels are oriented on the greenhouse with the long dimension

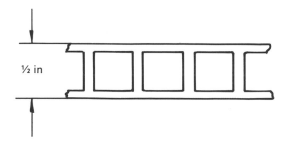

½ in

Cross section (approx. life size) of portion of a Rohaglas acrylic double-glazed panel

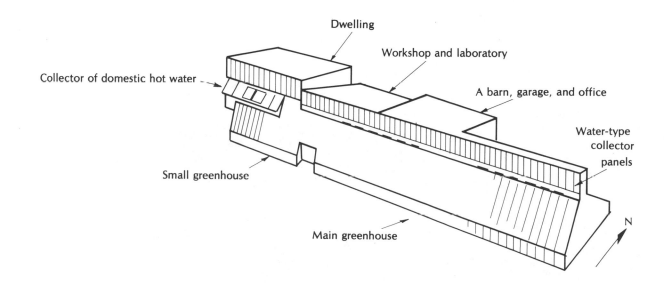

Dwelling

Workshop and laboratory

Collector of domestic hot water

A barn, garage, and office

Water-type collector panels

Small greenhouse

Main greenhouse

N

(and the partitions) running up and down the sloping array. Below this array there is a 100-foot-long row of 3-foot-high vertical windows double-glazed with glass. These windows admit a moderate amount of radiation.

Active solar heating system for dwelling. A water-type collector is employed, having an area of 750 square feet gross (700 square feet net). The collector is mounted along the upper part of the south face of the building, and is vertical. Its dimensions are 108 by 7 feet. It consists of thirty-six Sunworks panels, each 7 by 3 feet. Each contains a .010-inch-thick copper absorber sheet with a selective black coating (a/e = 0.9/0.1). Copper tubes ¼ inch in diameter and 6 inches apart on centers are soldered to the sheet. The glazing is single and consists of 3/16-inch tempered glass. The panel backing includes 2½ inches of fiberglass. The coolant (water, with no antifreeze) is circulated through the panels in parallel, at a combined flowrate of 12 gpm, by a ½-hp centrifugal pump. Hot water from the collector flows directly into the storage system; there is no heat exchanger. The storage system includes 17,000 U.S. gallons of water, which is contained in three tanks situated in the basement beneath the kitchen-dining-living area. The tanks are of different capacities: 2000 gallons, 5200 gallons, and 9700 gallons. Each is a rectangular poured-concrete tank and is insulated on the inside with 4 inches of sprayed urethane foam that has a waterproof butyl rubber coating. The routing of the hot water from the collector is arranged so as to give priority to keeping the smallest of the tanks hot. When the rooms need heat, a conventional forced-hot-air system is turned on.

Temperature stabilizing system for main greenhouse. This includes, besides the large amount of earth involved and the set of thirty water filled cylindrical tanks, a bin-of-stones situated on the north side of the building, beneath the barn and garage area. The bin, made of poured concrete and insulated on the outside with 4 inches of Styrofoam, is 22 by 20 feet by 6 feet high. It contains 118 cubic yards (about 200 tons) of 4-inch-diameter stones. The stones receive heat on sunny days when hot air accumulates in the upper part of the greenhouse space; hot air from this space is drawn into a duct and transported to the plenum beneath the mass of stones. The air passes upward through the mass of stones, warming them, and is then ducted to registers along the south edge of the greenhouse. The blower used is rated at 1½ hp. It is adjacent to the south side of the bin. When the greenhouse threatens to become excessively cold, the blower is again operated in the same manner—except that now, because the stones are hotter than the air in the upper part of the greenhouse, the overall result is transport of heat from the bin to the greenhouse. If the supply of heat in the stones is exhausted and the greenhouse has cooled down to about 50°F, additional heat can be introduced to the duct-and-blower system by means of a coil situated in the duct, close to the blower; the coil is supplied with water from one of the three tanks beneath the dwelling. Even if this water is only 70°F, it can supply a useful amount of heat, which may be regarded as being practically "free."

Percent Solar Heated Dwelling: 60% (predicted). Greenhouse: near 100% (actual). (Note The air in the greenhouse is permitted—in coldest weather—to cool down to about 50°F; the plants and fish suffer no harm.)

Auxiliary Heat Source High-capacity (200,000 Btu/hr) H. S. Tarm wood-burning boiler, situated in the workshop. This heats the water in the adjacent set of three large rectangular tanks. (Note: Inasmuch as hot water from one of these tanks can be supplied to the coil associated with the blower serving the greenhouse, this furnace may be regarded as indirect backup for the greenhouse.) Also there is a small wood-burning stove (Jøtul) in the dwelling.

Domestic Hot Water This is solar-heated by a small, water-type solar heating system. The collector employs several kinds of panels (for comparison). Their collective area is about 120 square feet. They are situated just outside the south side of the second story of the dwelling and slope 60°. Hot water from the collector circulates, by gravity convection, to a tank situated in the attic space nearby. The tank is cylindrical and horizontal and has a capacity of 100 gallons. The coolant used is water, with no antifreeze; when water in the collector threatens to become very cold, it is drained and is later put to use in the greenhouse. In winter, use is made of a backup electrical heater.

Cooling in Summer There is no formal cooling. Much use is made of ventilation, via (a) twelve vents situated above the main window ar-

ray of the greenhouse and (b) the 100-foot long row of openable windows below that array. The large mass of earth involved and the large amount of water help reduce the rise in temperature of the air.

Problems and Modifications   During start-up of operation, one differential control performed incorrectly, with resulting loss of much heat. The designers believe that the heating of the greenhouse may be significantly improved by installation of appropriate curtains to be closed on cold nights.

General concept: New Alchemy Institute. *Architect and solar designer:* Solsearch Architects (David Bergmark, Ole Hammarlund, Paul Scharf). *Adviser:* Advanced Concepts, Environment Canada. *Land:* Rented from Government of Prince Edward Island. *Funding:* $354,000 provided by Canadian Ministry of State for Urban Affairs. *Owner and occupant:* New Alchemy Institute, P.E.I., Inc.

Ayer's Cliff   45°N
(15 mi N of Vermont
border)
On Creary Road, 3 mi
NE of center of
Ayer's Cliff

> **Building:** 1½-story, 1200 sq. ft.
> **Collector:** 600 sq. ft., air type
> **Storage:** 12 tons stones, 1400 lbs. water
> **% Solar-heated:** 60 (predicted)

Exceptionally well-insulated house, partly recessed into the earth, employing unusually large air-type collector and providing 60 percent solar heating at a 9000-degree-day location

Building   This two-bedroom wood-frame house has an overall length of 50 feet and a maximum width of 24 feet. The two-story part of the house is at the east end. The lower story of this part, containing kitchen-dining room and bathroom, is partly below grade. The upper story includes two bedrooms. The one-story part of the house includes the living room. There is no basement, but there is a 4-foot crawl space beneath the one-story part of the house. There is no attic and no garage. The floor of the two-story part, and likewise the floor of the crawl space, is a 3-inch concrete slab that rests on 6 inches of sand. The concrete-block foundation walls are insulated externally with 2 inches of Styrofoam and 1 inch of sprayed urethane foam. The latter provides airtightness at the sills. The walls include 2½ inches of

urethane foam. The roof includes 9½ inches of fiberglass and has an R-value of 28. About 60 percent of the 215-square-foot window area is on the north side. All windows are double-glazed. Many precautions have been taken to minimize air leaks. The building faces exactly south.

Collection The 50-by-12-foot air-type collector, on the south roof, slopes 60 degrees. The heart of the collector is a 0.025-inch aluminum sheet that has a factory-applied nonselective black coating. A total of eighteen such sheets, each 8 by 4 feet, are used. The glazing is single and consists of a .040-inch-thick sheet of Excelite (fiberglass-reinforced polyester with a Tedlar coating). Each sheet is 12 feet long by 2 feet wide, and runs vertically. It is secured by vertical cedar battens to underlying vertical cedar supports. Silicone sealant is used. Each sheet is concave toward the south, with the axis of curvature vertical; the sheet is dished 3 inches. The purpose of the dishing is to accommodate thermal expansion. Also, it is said to improve the appearance of the collector. The air in the 1-to-4-inch space between glazing and black absorbing sheet is stagnant. The backing of the collector includes a waterproof composition board and 6 inches of fiberglass. Between the backing and the black absorbing sheet there is a 1 5/8-inch airspace defined by equally spaced horizontal wooden strips. In the six horizontal channels defined by these strips, air is driven westward by a blower situated in the above-mentioned crawl space. At the east and west ends of the collector there are vertical headers connected (by insulated ducts 18 by 12 inches in cross section) to the storage system. The blower, which has a maximum power of ½ hp and a maximum flowrate of 3000 cfm, is controlled automatically so as to provide whatever flowrate is most appropriate to the level of irradiation of the collector and the temperature of the storage system.

Storage This two-component system includes 1400 pounds of water in 350 half-gallon plastic bottles and 12 tons of 1½-inch-diameter

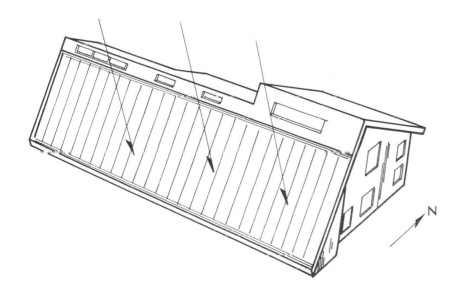

stones in a rectangular bin 16 by 8 feet by 4 feet high. The bin is in the south central part of the crawl space. It is insulated with 4 to 6 inches of foam. The air-handler includes, besides the above-mentioned blower, four dampers, a small electric motor for operating the dampers, and an automatic control system. There are three modes of air circulation: (1) collector air circulated through storage bin, (2) collector air circulated directly through rooms, (3) room air circulated through bin. In modes 1 and 3 the directions of airflow through the bin are opposite.

Auxiliary Heat Source   5 kw in-duct electrical heater. Also a small wood-burning stove.

Domestic Hot Water   This is preheated in a 120-gallon horizontal cylindrical tank within the storage bin. The tank is of glass-lined steel and is not insulated. In summer, such preheating continues: the collector is kept operating and the storage bin is kept hot.

Cooling in Summer   None needed, none provided.

Solar designer, builder, owner: Nick Nicholson. *Occupant:* Bruce Davidson. *Cost of house and solar heating system:* $38,000.

Addresses of Some of the Main Solar Designers and Solar Architects Mentioned

Warning: This is a selective listing, and the selection was made somewhat arbitrarily. Most of the solar designers, solar architects, and solar engineers mentioned in this book are included. But most of the persons, companies, and agencies involved in planning and constructing the 120 solar-heated buildings have been omitted. Some of the addresses listed may be incorrect; small companies often change their names and change their addresses, and individual solar designers and solar architects often change their affiliations and addresses.

Abraham, Kent: See Arthur Cotton Moore Associates

Acker, Gregory: see Living Systems

Acorn Structures, Inc., Box 127, Concord, MA 01742; also Acton, MA 01740.
 Bemis, John

Akehurst and Sun, 4138 Joppa Rd., Baltimore, MD 21236.
 Akehurst, Patrick F.

Alan Lower and Associates, Suite 900, National Foundation Life Building, 3535 NW 58 St., Oklahoma City, OK 73112.

Albuquerque Western Solar Industries, Inc., 612 Comanche, NE, Albuquerque, NM 87107.

Anderson, Bruce: see Total Environmental Action Inc.

Architects Taos: see The Architects Taos

Arthur Cotton Moore Associates, 1214 28 St. NW, Washington, DC 20007
 Abraham, Kent R.

Balcomb, J. Douglas: see University of California

Basford, Robert C.: see United States National Aeronautics and Space Administration

Banwell, White, and Arnold, Architects, Inc., 2 W. Wheelock St., Hanover, NH 03755.
 White, C. Stuart

Barber, Everett M., Jr.: see Enthone, Inc.

Benton, Crisp: see Massachusetts Institute of Technology.

Bier, James, Box 163, Chemistry Department, Ferrum College, Ferrum, VA 24088; also Rt. 2, Box 35, Ferrum, VA 24088.

Bishoprick, William: see Payne, Settecase, Smith and Partners

Blue Skies Radiant Homes, 40819 Park Ave., Hemet, CA 92343.
 Buckmaster, Warren D.

Boleyn, Douglas, 17610 Springhill Place, Gladstone, OR 97027.

Booth, Don: see Community Builders

Bridgers and Paxton, Consulting Engineers, Inc., 213 Truman St. NE, Albuquerque, NM 87108.
 Bridgers, Frank

Brown, Nelson: see Solar Technology, Inc.

Buckmaster, Warren D.: see Blue Skies Radiant Homes

Burman, James, Bay Rd., Hartford County, MD

Burt, Hill, Kosar, Rittelmann Associates, 400 Morgan Center, Butler, PA 16001.
 Rittelmann, P. Richard

Caivano, Roc, Russell Farm Rd., Bar Harbor, ME 04609.

Carleton Granbery Associates, 111 Old Quarry, Guilford, CT 06437.
 Granbery, Carleton

Carolina Solar Equipment Co., Inc., 301 N. Main St., PO Box 2068, Salisbury, NC 28144.
 Fisher, Daniel L.
 McCubbins, Benjamin D.

Carr, Laurence H., See Canyon Rd., Rt. 1, Box 170, San Luis Obispo, CA 93401.

Chalom, Mark: see Sundwelling Demonstration Center

CHI Housing Inc., PO Box 566, 68 St. Main St., Hanover, NH 03755.
 Carroll, John M.

Clark Enerson Partners, 1515 Sharp Building, Lincoln, NE 68508.
 Thomsen, Charles L.

Clark, LeRoy R.: see George C. Field Co.

Community Builders, Canterbury, NH 03224
 Booth, Don.

Costello, James M.: see Tritec Solar Industries

Crain, H. E., Inc., Flagstaff, AZ 86001.

Crosley, Mark, Box 36, St. Johnsbury, VT 05819.

Crowther Solar Group, 310 Steele St., Denver, CO 80206.
 Crowther, Richard L.
 Frey, Donald J.

Cuppett, H. Richard: see Forrest Coile and Associates.

Davies, Wolf, and Bibbins, Inc., 14 Arrow St., Cambridge, MA 02138.

Diamond, Edward: see Grumman Aerospace Corp.

DiBlazio, D.: see Forrest Coile and Associates

Dickinson, Spencer: see Solar Homes, Inc.

van Dresser, Peter, 634 Garcia St., Santa Fe, NM 87501. See also Sundwelling Demonstration Center

Dubin-Bloome Associates, 42 W 39 St., New York, NY 10018.
 Dubin, Fred S.

Duncan, Richard T.: see Westinghouse Electric Corp.

E-M Architects, 90 Low St., Concord, NH 03301
 Ellis, Bruce C.

Edmisten, John, School of Architecture and Design, California Polytechnic State University at San Luis Obispo, San Luis Obispo, CA 93401.

Ellis, Bruce: see E-M Architects

Enthone, Incorporated
 Sunworks Division, PO Box 1004, New Haven, CT 06508; also 669 Boston Post Road, Guilford, CT 06437.
 Barber, Everett M., Jr.

Field Co.: see George C. Field Co.

Fisher, Daniel L.: see Carolina Solar Equipment Co., Inc.

Forrest Coile and Associates, 11721 Jefferson Ave., Newport News, VA 23606.
 Cuppett, H. Richard.
 DiBlazio, D.

Frey, Donald J.: see Crowther Solar Group.

Frissora, J. R.: see Owens-Illinois, Inc.

George C. Field Co., Box 489, 107 Bradley Rd., Madison, CT 06443.
 Clark, LeRoy R.

Georgia Institute of Technology, Atlanta, GA 30332.
 Williams, J. Richard.

283

Shapiro, Andrew M, National Center for Appropriate Technology, PO Box 3838. Butte, MT 59701.

Shelton, Jay, 41 Porter St., Williamstown, MA 01267.

Shenandoah Development, Inc., PO Box 1157, Shenandoah, GA 30265.
    Moore, Ray.

Shore, Ron, PO Box 130, Snowmass, CO 81654.

Sky Therm Processing and Engineering, 2424 Wilshire Blvd., Los Angeles, CA 90057.
    Hay, Harold R.

Solar Applications and Research Ltd., 3356 W. 13th Ave., Vancouver, BC, Canada.
    Mattock, Christopher, 1729 Trafalgar St., Vancouver, BC, Canada.
    Summerlin, Paul.

Solar Design Associates, PO Box 153, Sharon, MA 02067.
    Strong, Steven J.

Solar Heat Corp., 108 Summer St., Arlington, MA 02174.
    Hyman, Mark.

Solar Homes, Inc., 2707 South County Trail, East Greenwich, RI 02818.
    Dickinson, Spencer.

Solar Room Co., Box 1377, Taos, NM 87571.
    Kenin, Stephen R.

Solar Structures Inc., 7 Sundance Rd., LaGrangeville, NY 12540.
    Wenning, Harry.

Solar Technology, Inc., 119 North Center St., Statesville, NC 28677.
    Brown, Nelson, 5730 Holiday Rd., Buford, GA 30518.

Solaron Corp., 300 Galleria Tower, 720 S. Colorado Blvd., Denver, CO 80222.

Solarsystems Industries, Ltd.: see Hoflar Industries, Ltd.

Sonnewald Service, RD 1, Box 1508, Spring Grove, PA 19362.
    Lefever, Harold R.

Speiser, Kenneth: see Grumman Aerospace Corp.

State University of New York at Albany, Atmospheric Sciences Research Center.
    Healey, James, Earth Sciences Building, Room 324, 1400 Washington Ave., Albany, NY 12222.

Strong, Steven J.: see Solar Design Associates.

Summerlin, Paul: see Solar Applications and Research Ltd.

Sun Harvester Corp., 729 Seventh Ave., New York, NY 10019.
    Price, Travis.

Sun Mountain Design, 107 Cienega St., Santa Fe, NM 87501.
    Haggard, Keith W.
    Lumpkins, William.

Sun Systems, Incorporated, PO Box 155, Eureka, IL 61530.
    Safdari, Y. B.

Sundwelling Demonstration Center, Ghost Ranch Conference Center, Abiquiu, NM 87510.
    Chalom, Mark.
    van Dresser, Peter.
    Haggard, Keith W.
    Lumpkins, William.
    Wright, David.

Suntek Research Associates, 500 Tamal Vista Blvd., #506, Corte Madera, Marin, CA 94925.

Terrien, George B., 165 Commercial St., Portland, ME 04111.

Terry, Karen, 636 Camino Lejo, Santa Fe, NM 87501.

The Architects Taos, Box 1884, Taos, NM.
    Haggard, Keith W.
    Mingenbach, William.

Thomason, Harry E.: see Thomason Solar Homes, Inc.

Thomason Solar Homes, Incorporated, 609 Cedar Ave., Oxon Hill, Fort Washington, MD 20022.
    Thomason, Harry E.

Thomsen, Charles L.: see Clark Enerson Partners.

Thornton, Richard, Garfield Road, Concord, MA 01742.

Total Environmental Action Inc., Box 47, Harrisville, NH 03450.
    Anderson, Bruce.
    Michal, Charles J., Jr.

Tritec Solar Industries, 711 Florida Rd., Durango, CO 81301.
    Costello, James M.

Tully, Gordon F.: see Massdesign Architects and Planners, Inc.

United States Department of Agriculture, Agricultural Research Service, Rural Housing Research Unit, Box 792, Clemson, SC 29631.
    Zornig, Harold F.

United States National Aeronautics and Space Administration, Langley Research Center, Hampton, VA 23665.
    Basford, Robert C.

University of California, Los Alamos Scientific Laboratory, PO Box 1663, Los Alamos, NM 87544.
    Balcomb, J. Douglas.
    Hedstrom, James C.
    Moore, Stanley W.

University of Dayton, Research Institute, Dayton, OH 45469.
    Whiteford, Dale H.

University of Maine, Orono, ME 04473.
    Hill, Richard C.

University of Texas at Arlington, Arlington, TX 76010.
    Hamilton, Todd, School of Architecture.
    Lawley, T. J.
    Lowery, G. W., Department of Mechanical Engineering.

van Dresser see "Dresser"

Wagoner, Robert G., Rt. 2, Sedona, AZ 86336.

Watson, Donald R., Box 401, Guilford, CT 06437.

Wells, Malcolm B., PO Box 1149, Brewster, MA 02631.

Wenning, Harry: see Solar Structures Inc.

Westinghouse Electric Corp.
    Solar Heating and Cooling Systems, Skyline Center, Suite 1307, 5205 Leesburg Pike, Falls Church, VA 22041.
    Duncan, Richard T.

Whedbee, John S., 294 Riverside Dr., New York, NY 10025.

White, C. Stuart: see Banwell, White, and Arnold, Architects, Inc.

Whitford, Dale H.: see University of Dayton.

Williams, J. Richard: see Georgia Institute of Technology.

Woodward, Richard M., 6A Chestnut Hill Rd., Black Mountain, NC 28711.

Wright, David, Box 49, The Sea Ranch, CA 95497. See also Sundwelling Demonstration Center.

Wright, Rodney: see Hawkweed Group, Ltd.

Yankee Barn Homes, Grantham, NH 03753.
    Gregor, Dick.

Zaugg Zaugg Architects, Inc., 60 South Main St., Mansfield, OH
44902.
    Zaugg, John H.
    Zaugg, Thomas Gene.

Zornig, Harold F.: see United States Department of Agriculture.

Photographic Credits

INDEX

292

NOTES